New Directions in Two-Year College Mathematics

New Directions in Two-Year College Mathematics

Proceedings of the Sloan Foundation Conference on Two-Year College Mathematics, held July 11–14 at Menlo College in Atherton, California

Edited by

Donald J. Albers Stephen B. Rodi
Ann E. Watkins

With 12 Illustrations

Springer-Verlag
New York Berlin Heidelberg Tokyo

Donald J. Albers
Department of Mathematics
Menlo College
Atherton, California 94025
U.S.A.

Stephen B. Rodi
Department of Mathematics
Austin Community College
Austin, Texas 78768
U.S.A

Ann E. Watkins
Department of Mathematics
Los Angeles Pierce College
Woodland Hills, California 91371
U.S.A.

AMS Classifications: 00A10, 00A99

Library of Congress Cataloging in Publication Data
Sloan Foundation Conference on Two-year College
Mathematics (1984 : Menlo College)
New directions in two-year college mathematics.
"Workshop held at Menlo College, Atherton, CA from
July 11–14, 1984, sponsored by the Alfred P. Sloan
Foundation"—T.p. verso.
Bibliography: p.
Includes index.
1. Mathematics—Study and teaching (Higher)—
United States—Congresses. 2. Community colleges—
United States—Curricula—Congresses. I. Albers,
Donald J. II. Rodi, S. B. (Stephen B.)
III. Watkins, A. E. (Ann E.) IV. Alfred P. Sloan
Foundation. V. Title.
QA13.S56 1984 510'.7'1173 85-8142

Conference held at Menlo College,
Atherton, CA from July 11–14, 1984.
Sponsored by the Alfred P. Sloan Foundation.

Printed in the United States of America.

Printed and bound by R.R. Donnelley & Sons, Harrisonburg, Virginia.

9 8 7 6 5 4 3 2 1

ISBN 0-387-96145-3 Springer-Verlag New York Berlin Heidelberg Tokyo
ISBN 3-540-96145-3 Springer-Verlag Berlin Heidelberg New York Tokyo

CONTENTS

Part 3. ENDANGERED CURRICULUM ELEMENTS

Part 4. NEW CURRICULA AND NEW TOOLS

Part 5. THE LEARNING OF MATHEMATICS

Part 6. FACULTY RENEWAL AND PROFESSIONALISM

Part 7. COORDINATING CURRICULUM CHANGE

INTRODUCTION

by

Donald J. Albers

INTRODUCTION

In July of 1984 the first national conference on mathematics education in two-year colleges was held at Menlo College. The conference was funded by the Alfred P. Sloan Foundation. Two-year colleges account for more than one-third of all undergraduate enrollments in mathematics, and more than one-half of all college freshmen are enrolled in two-year colleges. These two facts alone suggest the importance of mathematics education in two-year colleges, particularly to secondary schools, four-year colleges, and universities.

For a variety of reasons, four-year colleges and universities are relatively unaware of two-year colleges. Arthur Cohen, who was a participant at the "New Directions" conference warns: "Four-year colleges and universities ignore two-year colleges at their own peril." Ross Taylor, another conference participant, encouraged two-year college faculty to be ever mindful of their main source of students--secondary schools--and to work hard to strengthen their ties with them.

There are many other reasons why it was important to examine two-year college mathematics from a national perspective:

1. Over the last quarter century, no other sector of higher education has grown so rapidly as have two-year colleges. Their enrollments tripled in the 60's, doubled in the 70's, and continue to increase rapidly in the 80's.

2. Twenty-five years ago, two-year colleges accounted for only one-seventh of all undergraduate mathematics enrollments; today the fraction is more than one-third.

3. In 1960, most two-year colleges were called "junior colleges," reflecting a liberal-arts orientation and transfer function. In twenty-five years, the focus has greatly broadened. Today, only one third of two-year college students are enrolled in college transfer programs. Two-thirds of the students are enrolled in a host of vocational/technical programs, including data processing, dental hygiene, automotive mechanics, accounting, printing, bricklaying, carpentry, and police and fire science, to name a few.
4. Student populations in two-year colleges have changed significantly. In 1960, the population was made up primarily of 18- and 19-year old high school graduates, mostly single, mostly white, mostly males, and mostly attending on a full-time basis. Today, two-thirds of the students are over 21, one-third are married, some lack high school degrees, one-fourth are minority students, and more than one-half are women.
5. Technology has advanced very rapidly over the last quarter century. Powerful calculators

and computers are now readily available to students, and many believe that they should be playing more prominent roles in mathematics education.

6. New subject areas such as data analysis, discrete mathematics, and computer science have developed rapidly and have strong implications for mathematics education in two-year colleges.
7. A quarter century of developments in learning theory may have much to offer teachers of mathematics.

The foregoing served as the primary rationale for convening a working conference on mathematics in two-year colleges. Over the past decade, I have studied two-year colleges and have often felt the need for a national conference to sort through some of the problems of mathematics education. In 1983, a proposal for such a conference was sent to Dr. Stephen Maurer of the Alfred P. Sloan Foundation. He championed the conference idea, and the proposal was funded.

Several individuals, too numerous to name, helped in identifying conference participants. The conferees assembled for four days of work on the campus of Menlo College in July of 1984. Twenty-two papers were presented and discussed. In addition to the papers, this book contains edited discussions of the papers and "Recommendations of the Conference: New Directions in

Two-Year college Mathematics."

The success of the conference owes much to the conferees; to Joan Bailey, Administrative Assistant for the Conference; to Michael W. Larkin, Logistics Coordinator; to President Richard F. O'Brien, to Provost R. Eugene Bales, and Dean Joseph Zikmund II, all of Menlo College; and to my wife Geri and daughter Lisa.

Donald J. Albers

Chairman of the Conference

References:

(1) James T. Fey, Donald J. Albers, Wendell Fleming. Undergraduate Mathematical Sciences in Universities, Four-Year Colleges, and Two-Year Colleges, 1980-81. Conference Board of the Mathematical Sciences, Washington, 1981.

(2) Robert McKelvey, Donald J. Albers, Shlomo Libeskind, and Don Loftsgaarden. An Inquiry into the Graduate Training Needs of Two-Year College Teachers of Mathematics. Rocky Mountain Mathematics Consortium, University of Montana, Missoula, 1979. (A long summary of this report appeared in the March 1979 issue of the TWO-YEAR COLLEGE MATHEMATICS JOURNAL.)

(3) Donald J. Albers. A Faculty in Limbo, an essay in MATHEMATICS TOMORROW, Springer-Verlag, New York, 1981.

(4) Fontelle Gilbert. Fact Sheets on Two-Year Colleges. American Association of Community and Junior Colleges, Washington, 1983.

(5) Miriam M. Beckwith. Science Education in Two-Year Colleges: Mathematics. ERIC Clearinghouse for Junior Colleges, UCLA, 1980.

FOREWORD

by

Amber Steinmetz

FOREWORD

This book records the Proceedings of a working conference on mathematics education in two-year colleges held on July 11-14, 1984, at Menlo College in Atherton, California.

During the past two years numerous reports have indicated the need for changes within the educational system in the United States. The Conference Board of the Mathematical Sciences (CBMS), recognizing the need for change, has asked the basic question: "What is fundamental to mathematics education, and what is not?" Conferences have been held for the K-12 population, a conference was held at Williams College on the first two years of college mathematics, but this is the first conference devoted to the unique problems of two-year college mathematics.

Over the last quarter century, the objectives of two-year colleges have gradually shifted from primarily serving the needs of students who wish to transfer to four-year colleges to serving the needs of students who desire technical and job-related training. During this same period, rapidly advancing technology has made calculators and computers accessible to many students and faculty, and has greatly influenced the teaching of mathematics. To consider these needed changes and to present them in the form of recommendations to mathematical communities throughout the United States, it was decided that conference participants should include representatives from two-year colleges, from vocational and professional organizations, as well as from secondary schools and universities. Participants prepared and distributed their papers to other participants before the opening of the conference and lead a discussion of their papers at the conference itself.

It was the unanimous decision of the participants that the recommendations, focusing on curriculum and the new technologies, faculty development and renewal, and collaborative efforts from all levels of mathematics education, should be forwarded to national organizations which support the mathematical sciences. If these organizations endorse the recommendations, ways and means for their implementation should be investigated.

Thanks are extended to Donald J. Albers, conference chairman and organizer, to Menlo College for providing the conference site, to the Sloan Foundation for its financial support, and to the many conference participants who took time from their busy summer schedules to work on this important project.

Amber Steinmetz

Recommendations from the Conference

NEW DIRECTIONS IN TWO-YEAR COLLEGE MATHEMATICS

New Directions in Two-Year College Mathematics

Twenty-four leaders in American education met at Menlo College in Atherton, California in July of 1984 in order to discuss the state of mathematics in two-year colleges. Twenty-two papers were presented and discussed. The main topics addressed by conference participants included the following: a case for curriculum change, technical mathematics, the influence of new technologies on the learning of mathematics, faculty renewal, and collaboration with secondary schools, colleges, and universities. Presented here are the recommendations that emerged from the conference.

The Curriculum and the New Technologies

Mathematics courses in two-year colleges should be of immediate use to students and not be seen merely as preparation for distant goals. The material should be connected to real life. Those parts of the subject of broadest usefulness in problem solving should be emphasized.

Computers and other elements of information technology are changing in fundamental ways both the mathematics that it is important to learn (such as discrete mathematics, statistics, and technical mathematics) and the ways that mathematics, traditional and otherwise, can be learned. This will have profound and continuing implications for students, faculty, and the curriculum.

Successful integration of information technology into the curriculum requires immediate access to computers; thus all mathematics faculty must be provided by their institutions with appropriate computer equipment and support and with training opportunities in information technology and related mathematical subjects.

- Statistical literacy should be a fundamental goal of schooling. Basic mathematics courses should contain elementary statistical ideas and must prepare students for statistics as well as for calculus. Mathematics faculty should be trained in statistics.
- Geometric concepts, including the use of computer graphics, should be integrated into entry-level courses wherever appropriate.
- Entry-level courses for students who lack arithmetic skills should be organized around mathematical content new to these students (for example, vocational mathematics) and should utilize approaches different from what they may have experienced in school mathematics (for example, an approach emphasizing problem solving with calculators).
- Liberal arts mathematics (mathematics appreciation) is an important course and should integrate basic mathematical competencies including new technologies. These courses should be organized around the great ideas of mathematics and should offer a variety of topics for choice by instructors. Included in this choice should be topics from statistics, computing, and discrete mathematics.
- New mathematics curricula and approaches based on information technology (for example, spreadsheets, databases, and computer graphics) should be developed to serve the special vocational/technical and functional needs of the diverse two-year college student population.
- Planning should begin immediately to deal with the impact of symbolic manipulation software, graphics-based software, and the algorithmic point of view on the content of all mathematics courses.
- Two-year colleges must be prepared to meet the needs of entering students having a wide range of prior experience with computers.

Collaborative Efforts

Two-year colleges enroll more than half of all college freshmen and share common concerns about mathematics students with secondary schools, colleges, and universities. The success of our students will be enhanced by collaboration including the following general activities: development of policies and methods for cooperation by the principal mathematics organizations concerned with teaching at these levels (American Mathematical Association of Two-Year Colleges, Mathematical Association of America, National Council of Teachers of Mathematics); joint efforts by state and local mathematics organizations and institutions; personal initiatives on a professional and social basis by individual faculty at all levels to promote a coherent and appropriate mathematics curricula.

- **It is imperative to reduce the need for remediation in the colleges.** Two-year colleges along with other post-secondary institutions must work together with high schools to this end.
- Two-year college faculty in mathematics and in vocational/technical areas should meet within their own institutions and at regional levels to form better working relationships and to improve the quality of mathematics courses for vocational/technical students.
- Training programs should be instituted for mathematics faculty in order to increase skills in teaching technical mathematics. These programs will require collaboration and support by the private sector, two-year colleges, foundations, and professional organizations.
- High schools, four-year colleges and universities, and two-year colleges should establish faculty exchange programs for mutual benefit.

Faculty Development and Renewal

Faculty development and renewal are essential in maintaining the vigor of instruction and the quality of the curriculum. Faculty must commit themselves to continued and on-going professional growth and development. Thus, the educational community, industry, government, and foundations should provide adequate funding to ensure continued and frequent opportunities for faculty development. Professional organizations should provide a variety of activities and assistance for faculty growth. College administrators should encourage, support, acknowledge, and reward renewal efforts by faculty.

- Faculty must foster their own continued growth. Two-year colleges should view these efforts of faculty as essential and support them financially. Such growth activities include regularly teaching a variety of courses, reading, writing papers, participating in seminars and in professional organizations, and learning new material through workshops and courses.
- Two-year colleges and their supporting bodies should view employment compensation and status more flexibly in order to encourage and reward professional excellence.
- The private sector and two-year colleges should facilitate faculty participation in industry, including short-term work experiences, as a valuable form of development.
- Universities should provide graduate programs appropriate for two-year college mathematics faculty.

Conference Participants

Donald J. Albers
Menlo College
Atherton, California
(Chairman of the Conference)

Bettye Anne Case
Tallahassee Community College and
Florida State University
Tallahassee, Florida

Arthur M. Cohen
ERIC Clearinghouse for
Junior Colleges, UCLA
Los Angeles, California

Larry A. Curnutt
Bellevue Community College
Bellevue, Washington

Ronald M. Davis
Northern Virginia Community College
Alexandria, Virginia

Wade Ellis
West Valley College
Saratoga, California

Ben Fusaro
Salisbury State College
Salisbury, Maryland

Sheldon Gordon
Suffolk County Community College
Selden, New York

James Kaput
Southeastern Massachusetts University
North Dartmouth, Massachusetts

Joan R. Leitzel
Ohio State University
Columbus, Ohio

Calvin T. Long
Washington State University
Pullman, Washington

Stephen B. Maurer
Alfred P. Sloan Foundation
New York, New York and
Swarthmore College
Swarthmore, Pennsylvania

Warren Page
New York City Technical College
Brooklyn, New York

Peter Renz
W. H. Freeman and Company
New York, New York and
Bard College
Annandale-on-Hudson, New York

Stephen Rodi
Austin Community College
Austin, Texas

Karen Tobey Sharp
Charles Stewart Mott Community Colleg
Flint, Michigan

Keith Shuert
Oakland Community College
Auburn Heights, Michigan

Karl Smith
Santa Rosa Junior College
Santa Rosa, California

Amber Steinmetz
Santa Rosa Junior College
Santa Rosa, California

Ross Taylor
Minneapolis Public Schools
Minneapolis, Minnesota

Alan Tucker
SUNY at Stony Brook
Stony Brook, New York

William Warren
Southern Maine Vocational
Technical Institute
South Portland, Maine and
Council for Occupational Education
Annandale, Virginia

Allyn Washington
Dutchess Community College
Poughkeepsie, New York

Ann Watkins
Los Angeles Pierce College
Woodland Hills, California

Geoffrey Akst of the Borough of Manhattan Community College prepared a paper for the Conference but was unable to attend due to illness. Jerome A. Goldstein, Tulane University, coauthored one of the Conference papers.

Part 1.

A CASE FOR CURRICULUM CHANGE

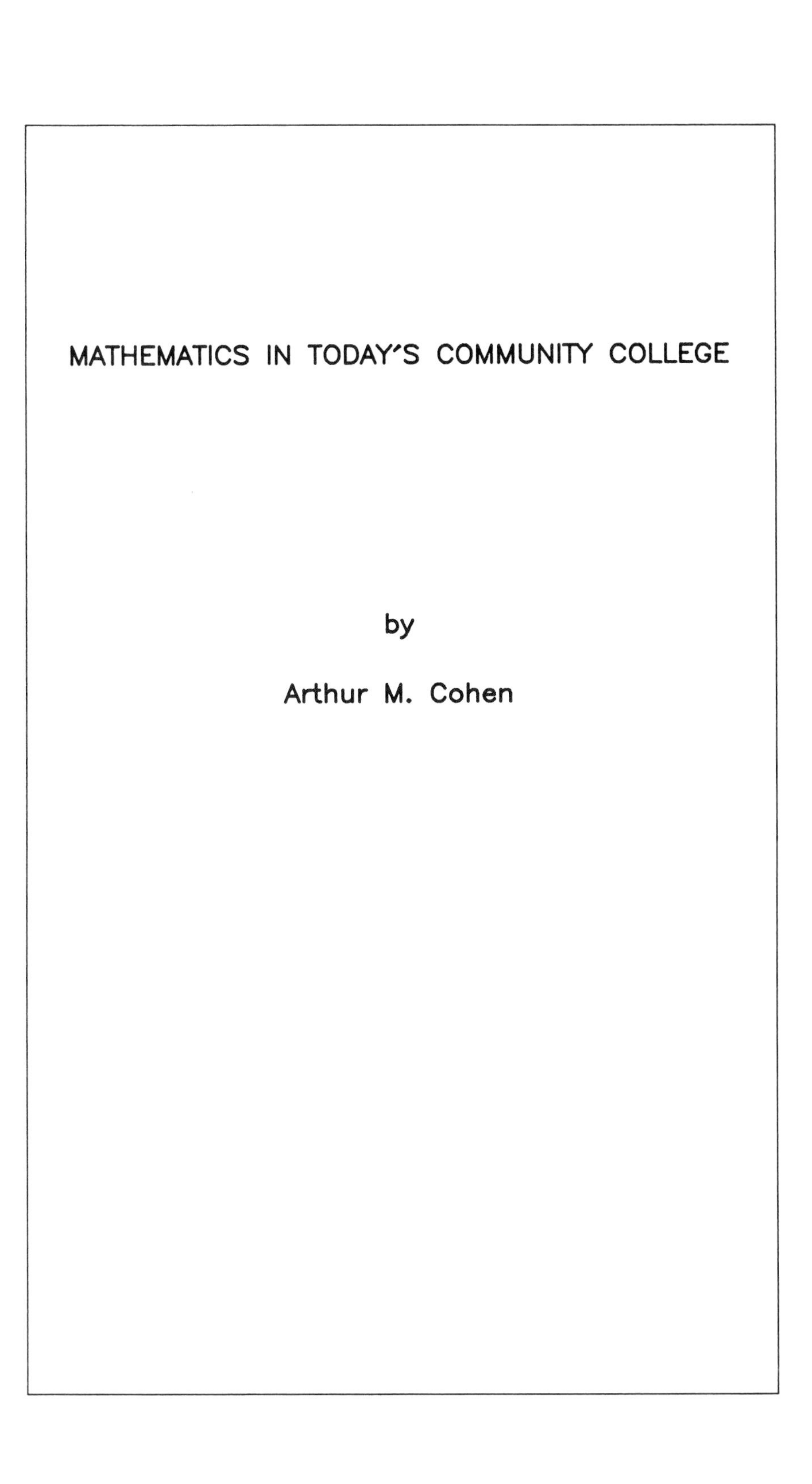

MATHEMATICS IN TODAY'S COMMUNITY COLLEGE

by

Arthur M. Cohen

SUMMARY

The peculiarities of community college mission, students, and staff affect the teaching of mathematics in those institutio Although most of the colleges offer introductory, intermediate, and advanced mathematics courses, well over half the enrollments are in courses teaching pre-algebra, introductory algebra, geometry, college algebra and trigonometry, and/or at the developmental or remedial level. Mathematics Learning Centers are well-established to teach developmental courses as well as to offer tutorial and supplemental instruction to students in the other courses. The faculty teaching the developmental courses make wide use of tutors and reproducible aids to instruction and seem to follow precepts stemming from the discipline of instruction more closely than their counterparts in the advanced classes. All the instructors must be concerned with teaching mathematics to students in technological programs as well as to those planning on university transfer and those seeking only skills qualifying them for immediate employment. The problem is confounded by the fact that any student may be at once or successively a member of all three groups.

Mathematics in Today's Community College

Arthur M. Cohen

In the past 20 years the American community college has been converted from a university preparation institution to a career and compensatory education center. Whereas the colleges formerly emphasized transfer studies and took pride in the number of students they attracted from high school and sent on to the universities, more recently they have developed a variety of career-oriented programs and have been forced to modify all their programs to accommodate the marginally literate students emanating from the secondary schools. The conversion was not absolute --- many students still use the colleges as a point of entry to baccalaureate studies -- but it did mark a shift in the institutions' central purpose.

This modification of function has affected the institutions' people, processes, and programs. No aspect has been immune. Faculty have had to change teaching practices; the very number of pages they can expect students to read has plummeted. Career guidance has supplanted academic advisement. In most colleges, ten sections of remedial reading or writing are offered for every one section of English or American literature. And the part-time adult student seeking a course in job upgrading has become a familiar figure on campus.

Some analysts applaud the colleges' break with the tradition of the higher learning, viewing the changes as necessary for an institution that would serve a mass public. Others deplore the turn away from collegiate studies, feeling that the colleges thereby do a disservice to the students who look to them as the place to begin an education that will bring higher social and fiscal rewards than those otherwise attainable. Regardless of one's position on the broader aspects of the changes, they should be examined by everyone involved with the colleges because they affect all, not least the instructors of mathematics, to whom this paper is addressed

STUDENTS

The 1,250 community colleges of America enroll more than one-third of all people engaged in formal post-secondary education. In some states more than 80% of the people who begin college begin in a community college. This growth was occasioned by the increased percentage of people seeking post-secondary studies. Whereas at the turn of the century, less than 2% of the traditional college-age group was in college, by 1981 42% of the 18 to 24 year-olds were attending. The community college contributed to this growth by making access easier for people who in an earlier era would not have consider college-going: minority group and low-ability students, studen from low-income families, students who wished to enter the workplace but found they needed an additional year or two of schooling in order to gain job preparation and a certificate enabling them to apply for employment in certain fields. The expansion of career education programs reflects this latter trend: in 1960 one-fourth of the community college students were enrolled in career programs; in 1970 the colleges awarded 43%, and in 1980, 62% of their associate in arts and sciences degrees to career program graduates.

The colleges grew also by encouraging part-time attendance In 1968 they enrolled 1.9 million degree credit students, 47% of whom were attending part-time. In 1982 they enrolled 4.9 million students, with 63% attending part-time. With the exception of New York and North Carolina, in the fourteen states with community college enrollments greater than 50,000, part-time students outnumbered the full-timers.

Students of lower ability have swelled community college enrollments. Most American colleges have some type of selecti in admissions but more than half the community colleges allow students to attend if they are of a minimum age (usually 18) and/or they present a high school diploma. Only around one-fourth of them ask the students to provide ability test scores and few, if any, use the students' high school grade point average as a criterion for admission. This has resulted in a high proportion of students with poor prior academic records attending community colleges. Whereas 62% of the full-time students entering all post-secondary institutions in 1983 were from the top 40% of their high school class, only 47% of that group entered a community college.

The colleges have also attracted sizeable proportions of the ethnic minorities and similarly high proportions of students from low-income families. By 1980 the colleges were enrolling nearly 40% of the ethnic minority students involved in American higher education, and more than half the minorities beginning college began in a community college. The distribution of family income similarly showed a tilt toward low-income students: 54% of all first-time, full-time students entering college were from families with an annual income of less than $35,000, but 74% of the community college matriculants fell into that category (Astin and others, 1982).

These characteristics of community colleges--lower ability students, students wishing to find employment after only a short period of study, part-time attendees, students who come from families without a tradition of college going--have marked both the curriculum and the faculty. The curriculum in most traditionally organized schools has the appearance of a ladder. Each course builds on the one prerequisite to it and all are presumed to be part of a sequence resulting in a degree or diploma that evidences certain knowledge and competencies. The community colleges have broken away from that sequence. By far the majority of their enrollments are in introductory courses for which there is no prerequisite: English 1, Psychology 1, History 1, and so on. Relatively few students are enrolled in advanced or second level courses. As an example, an enrollment count undertaken in 1980 in five of the largest community college districts in the country revealed the following patterns: in the humanities there were no remedial courses, but 85% of the enrollment was in introductory courses or courses for which there was no prerequisite. The pattern for other disciplines was comparable: social sciences, 79% in introductory courses; sciences, 5% remedial, 65% introductory; mathematics, 60% remedial, 25% introductory; composition, 35% remedial, 50% introductory. (The only exception was in the fine and performing arts which had no remedial classes, 60% introductory, 40% advanced). The curricu in those institutions had become grade 13 plus remedial studies with only a nod toward advanced work. These patterns are reflected in the transfer rate which shows less than 5% of the students entering community colleges nationwide completing two years there and then transferring to a senior institution. What transfer does occur often takes place after the first year when the student, having made a satisfactory record in a set of introductory courses, transfers to a neighboring senior institution for the sophomore and subsequent years. (Cohen and Brawer, 1982).

CURRICULUM

The effect of these shifts on mathematics can be traced because the literature about mathematics in the community colleges is better developed than the literature regarding othe academic fields taught in those institutions. Mathematics is a well-defined curriculum area and it accounts for a higher percentage of community college effort than any discipline othe than English composition, reading, and literature. A National Science Foundation-sponsored study done in 1969 found mathemati accounting for 24% of the two-year college enrollment in scienc (Beckwith, 1980, page 5). Among the associations directing papers toward people teaching and planning mathematics in the community colleges is the Committee on the Undergraduate Progra in Mathematics, a standing committee of the Mathematics Associa of America. Major sources of information also include such journals as the *Mathematics Association of Two-Year Colleges Journal* and the *Two-Year College Mathematics Journal*.

Many of the reports coming from the Committee on the Undergraduate Program in Mathematics in the 1960s and 1970s related to what the group perceived as deficiencies in the curriculum in two-year colleges. They reported on the need for additional courses in calculus, linear algebra, and other advanced levels of mathematics and on the preparation of instructors in mathematics. However, more recently, such conce for the upper reaches of mathematics in two-year colleges has given way to consideration of mathematics for general education or for the severely underprepared student. The monograph entitled *Science Education In Two-Year Colleges: Mathematics* (Beckwith, 1980) reviewed several studies of mathematics in general education, including one done in 1974 that found two-thirds of the mathematics departments offering a course specifically designated for general education.

Studies of remedial mathematics have included nationwide surveys on teaching techniques, faculty qualifications, time spent on various aspects of mathematics, and related issues of general concern. Chang (1983) found elementary algebra offered as part of the content in 82% of the remedial courses, arithmetic in 68%, and intermediate algebra in 53%. Most of the colleges surveyed did not offer credit for remedial courses. Nearly all the colleges provided tutorial services for the students and, in just under half the cases, the remedial students finished their developmental programs within one semester; just over half of them went on to complete at least one college-level mathematics course. Beckwith reported a nationwide study, done by the Center for the Study of Community Colleges under National Science Foundation auspices, which revealed that 97% of the colleges offered introductory and intermediate mathematics, 86% offered some advanced study, and 67% offered applied or technology related mathematics. Well over half the course offerings were at the introductory or intermediate level, including courses in pre-algebra, introductory algebra, geometry, intermediate algebra and trigonometry, college algebra and trigonometry, and all developmental or remedial courses.

The literature on remedial education in mathematics revolves around questions of definition, placement of students, granting of credit, course content, and effective instructional practices. Beckwith reviewed a 1975 study which found that nearly all two-year colleges offered courses in developmental mathematics. At that time less than half the colleges required a placement examination and, in those that did, few mandated that students take a remedial course. That was the era when the students were to be given the right to fail, a procedure and value system that has changed notably in the past few years. Now, a rapidly increasing number of two-year colleges are returning to the 1950s pattern of mandated placement tests and course placements in English and mathematics.

Many reports from individual colleges are also available.

Developmental mathematics at Lower Columbia College (Washingto
was described in a report indicating the various ways that students might complete the mathematical requirements through courses offered in laboratory and classroom. As in many other colleges, the high attrition rates in developmental mathematic were combated by extensive placement testing, math anxiety workshops, and several levels of review courses (Crepin, 1981) Math avoidance was revealed in a study of students' course-taking patterns in a large urban community college district in which between 12 and 36 percent of the students who had not an interest in science-related careers had completed no mathem courses even after having completed more than thirty units of college work (Friedlander, 1981, Table 5).

Information about the operation of laboratories and cente that teach basic mathematics is prevalent. Mitchell (1980) reported on the operations of the Mathematics Center at Pima College, indicating that its 20 to 25 staff served 400 to 500 students. It had a lead faculty member or coordinator respons for employing and scheduling instructors, assistants, and tuto training the staff; establishing record keeping systems; developing curriculum; managing the preparation of materials; determining grades; and "serving as the instructor on duty several hours each week (p. 43)." The Center had a differenti staff with other fully certified instructors, clerical assista and peer tutors. Problems at the Center included computing faculty work load and training peer tutors. The Mathematics Learning Center at Cerritos College was examined from the standpoint of its financial base. The Center was cost-efficie because its course sections were large and because it employed paraprofessional personnel to maintain student records. The author recommended keeping the staff lean and insuring that students understand the Center's policies (Baley, 1981). Some of the reports of the mathematics centers and laboratories describe the special services offered for students with variou types of deficiencies in mathematics understanding (Habib, 19 Yawin, 1981; Rotman, 1982). Others focus more intently on the organization and operation of the laboratories themselves (Emerson, 1978; Fast, 1980; Palow, 1979).

STUDENT COMPETENCIES

Some work has been done recently on competencies expected of entering freshmen, with particular attention to tests that would place students in mathematics courses in which they had a reasonable chance of succeeding. Many of the questions swirling around student placement have to do with the relative merits of homemade placement tests and tests that are prepared and distributed by a national agency. Several articles on faculty dissatisfaction with nationally-normed tests have been published. Wood (1980) reports the switch from the ACT mathematics placement test to an instrument developed at the University of Houston, Downtown Campus, more than fifteen years ago. The college used that 50-minute test to shunt students scoring below seventy percent to a course reviewing algebra. Notable results in retention and achievement resulted.

A minimum level of mathematical literacy for all college-level students was specified in the 1981 annual report of the American Mathematical Association of Two-Year Colleges Developmental Mathematics Curriculum Committee (Dyer, 1981). In several states the community colleges have been cooperating with the universities and secondary schools in announcing competencies expected. The Academic Senates of the California community colleges, the California State University, and the University of California addressed a statement to parents, teachers, counselors, and administrators of high school students indicating the requisite competencies in algebra, arithmetic, geometry, and advanced mathematics that students should demonstrate in preparation for college (Statements..., 1982). Miami-Dade Community College (Florida) prepared a booklet for distribution to junior high school and high school students in its service area indicating expected competencies in mathematics and English usage. Community colleges elsewhere have undertaken similar projects.

Despite these pronouncements, the question of how much students know is studied only rarely. As a way of getting information about absolute levels of student knowledge, the National Assessment of Educational Progress has for more than fifteen years administered items in science, social science, and mathematics to samples of nine year-olds, thirteen year-olds, and seventeen year-olds. Building on this idea of assessing student knowledge in an absolute rather than a relati sense, the Center for the Study of Community Colleges designed an instrument to survey community college students' knowledge in the humanities, sciences, social sciences, mathematics, and English usage. Many of the items in this General Academic Assessment were provided by the National Assessment of Educatic Progress with other items borrowed from Educational Testing Service or provided by various community colleges. The test included 53 items in mathematics spread over five forms; the scores were converted to ten point scales. In addition the Center collected such background items about the students as age, number of college credits earned, occupational aspirations number of courses taken in each liberal arts area, self-assessr of academic skills, and primary reason for attending college.

The test was administered to 8,026 students in the commun: colleges of Los Angeles, Miami, St. Louis, and Chicago. The sample was generated by picking every Nth academic transfer-credit class from the 1983-1984 schedules. Excluded from sampling were remedial classes, occupational classes that did not carry transfer credit, and adult education or community service courses. The class section was used as the unit of sampling because it is the most feasible way of administering a survey to a random sample of community college students.

The results were as expected. Students who had completed more total college units, who anticipated receiving an associa degree by June 1984, who were in college to transfer and/or to prepare for a career in the advanced professions or technologi or who had completed more math courses scored higher. But the highest correlation of all was between the mathematics scores and the question, "Compared to other students at this college how would you rate your ability to use algebra to solve proble Students were given a choice of rating themselves, poor, fair, good, or excellent. Scores on the mathematics scale ranged from below four for the students who rated themselves as poor to above six for those who said they were excellent (Riley, 1984).

INSTRUCTION

The future of mathematics in the community colleges will

see more modification in instructional design than in course content. The computer and hand-held calculator have already changed patterns of drill and demonstration. Advances in those technologies will continue to make inroads on traditional instruction. Enrollments will remain high because mathematics is a service course to numerous technologies and career programs. A 1977-1978 survey found 449,000 students taking mathematics classes in community colleges nationwide. And that figure does not include those who were enrolled in remedial sections offered through mathematics laboratories. The number was the highest for enrollments in any academic area within the sciences, social sciences, or humanities. Contrast that figure with 335,000 enrollments in history, 255,000 in political science, and 225,000 in psychology. Or, put another way, contrast it with the 73,000 students studying chemistry or the 35,000 enrolled in physics courses. (Cohen and Brawer, 1982, p. 289)

Changes in the mode of presenting concepts and drill in mathematics depend in large measure on the way the instructors perceive their role. Education is a labor-intensive enterprise; will the faculty continue demanding small classes? Nationwide, classes in mathematics are smaller than classes in the other sciences and social sciences: 28 vs. 32, on average. Of the scheduled mathematics classes in six of the largest districts, 21% had fewer than 20 students; 56% between 20 and 30; and 23% more than 30. Yet 35% of the instructors said their classes would be more effective if they had fewer students.

Based on the literature there seems to be a split between instruction in remedial mathematics and college level mathematics as great as the gulf between the teaching of English composition and English literature. The publications on remedial mathematics speak of laboratory experiences, tests, grades, auto-instructional programs, and ways of staffing the laboratories to make them more efficient. The articles on college-level mathematics discuss games, proofs, problem-solving strategies, theorems, and the unfortunately labeled concept, math anxiety. This split is revealed in the articles carried in *The Two-Year College Mathematics Journal* in which, for the four years beginning in 1980, there were tips on teaching, mathematics concepts for classroom use, mathematics games, and a few articles on mathematics avoidance. There was an occasional article on

merging mathematics with other fields, as for example, "Integrat
Writing into the Mathematics Curriculum," (Goldberg, 1983), and
few reports of classroom experience, for example one involving
basic mathematics and women in which the investigator found that
women in all-female section of a basic algebra class did better
than women in mixed sex classes (Brunson, 1983). But most of
the papers were distinctly addressed to instructors of advanced
classes, whereas the ERIC system and the journals especially
slanted toward remedial studies in mathematics and English usage
carry papers describing the operations of learning laboratories.

The major differences between remedial and college-level
mathematics seem to be in the pattern of presentation (laborator
vs. classroom) and in the staffing (a lead instructor supervisin
a corps of aides vs. a lone instructor in a classroom). Further
differences appeared in the Center for the Study of Community
Colleges' national surveys. Instructors of remedial classes
indicated they spent less time in lecture than did instructors
of advanced classes (36% vs. 49%). They were less likely to
administer ABCDF grades (52% vs. 76%). As a group they tended
to be younger, with less teaching experience, but much more
likely to rely on tutors (60% vs. 47%) and paraprofessional
aides (27% vs. 12%) for teaching assistance and more likely to
use test-scoring facilities (23% vs. 13%)(Beckwith, 1980).
These differences suggest that the instructors of remedial
classes are perforce leading in the development of a profession
faculty peculiarly suited for the community college.

FACULTY

Community college instructors in general differ from their
senior institution counterparts in demographic characteristics,
attitudes, and values. The community college faculty teaching
transfer-credit courses typically hold the Master's degree, and
the instructors of occupational subjects tend to be certified
on the basis of their experience within the trades that they
teach. Members of both groups have relatively high teaching
loads, with the instructors of transfer courses teaching from
13 to 16 hours per week or four or five classes with around
30 students in each. The occupational program faculty often
teach longer hours since they are involved in clinics and
laboratories.

The faculty tend not to be members of academic disciplinary associations. As an example, less than 7% of the people teaching history belong to the American Historical Association and similar figures pertain for community college faculty membership in the American Philosophical Association, the American Sociological Association, and so on. Where associations have been formed with the intent particularly to involve community college instructors, the membership ratio is much higher. The Community College Humanities Association has developed into a thriving national group over the past five years and 17 percent of the full-time mathematics instructors are members of the American Mathematical Association of Two-Year Colleges.

Are the faculty satisfied with their working conditions? Until the 1960s the local secondary schools were the largest single source of community college instructors. For those who moved from a secondary school to a community college, faculty satisfaction was high because they had entered a higher status position and enjoyed a reduced teaching load. The less satisfied instructors tended to be the younger ones coming in directly from graduate school. Still, the faculty continually pleads for better qualified students, smaller classes, and more time off. In one large, urban community college district recently, the faculty bargaining unit negotiated a teaching load reduced from 15 to 12 hours per week. In return they relinquished all sabbatical leaves, instructional development grants, and travel funds. They saw lower teaching loads as more crucial to their professional well being and personal satisfaction than the perquisites that faculty historically have indicated as being essential for their professional currency.

This sheds light on the question of professional status. Some commentators have reasoned that the community college is best served by a group of instructors with minimal allegiance to a profession. They contend that professionalism leads to a form of cosmopolitanism that ill-suits a community-centered institution, that once faculty members find common cause with their counterparts in other institutions they lose their loyalty to their own colleges. This argument stems from a view of professionalism among university faculties that has ill-suited teaching in the senior institutions, where, as faculty allegiance turned more to research, scholarship, and academic disciplinary concerns, interest in teaching waned.

However, that argument suggests that a professionalized community college faculty would necessarily take a form similar to that taken by the university faculty. It need not. It could develop in a different direction entirely, tending neither toward the esoterica of the disciplines nor toward research and scholarship on disciplinary concerns. The community college faculty disciplinary affiliation is too weak, the institutions' demands for scholarship are practically non-existent, and the teaching loads are too heavy for that form of professionalism to occur.

A professionalized community college faculty might well organize itself around the discipline of instruction. The faculty is already engaged in course modification, the productio of reproducible teaching media, and a variety of related activities centered on translating knowledge into more understandable forms. They have had to take that direction because of the paucity of self-directed learners in their classrooms. A corps of professionalized instructors would well suit the community college. They could reform curriculum, devise entry and exit examinations for their students, manage groups of paraprofessional aides and instructional assistants, prepare reproducible instructional media, and exhibit the other essential components of people practicing the discipline of instruction. A professionalized community college instructor would be a manager of student learning. It is likely that the instructors most nearly acting like such practitioners currently are involved with learning laboratories that have taken over much of the remedial instruction.

Many two-year college instructors have so professionalized themselves, but that concentration on the discipline of instruct has not yet become the hallmark of a group of instructors sufficiently large to have it seem the central tendency of the 200,000 people teaching in the nation's community colleges. The relatively heavy teaching loads take their toll, and as long as instructors insist on smaller and fewer classes, instruction remains a labor-intensive, high-cost enterprise. Fortunately for the development of community college teaching as a professic many instructors in remedial mathematics have taken the lead in developing learning laboratories and in pursuing instruction through differentiated staffing.

Few mathematics teachers enjoy presiding over students doing drills in the classrooms, but neither should they want to turn the laboratories over to less qualified staff. They are faced with becoming managers of learning which involves them in designing placement tests, analyzing the results of instruction, defining objectives, monitoring procedures, reviewing programs, developing competency tests, and conducting follow-up studies of students who transfer or who enter the work place. They can do that or risk being shunted to the corners of the community college where they would find the occasional serious student of mathematics who would enter a career in one of the advanced technologies or who would transfer to the university and major in science. Too few of the community college students fit those categories to warrant a sizeable staff teaching mathematics beyond the level of college algebra.

The experience of other disciplines whose practitioners insisted on teaching only the majors in their area or the students who were serious about their studies is instructive. In some colleges the instructors in specialized areas in the humanities recognized that there were not enough serious students to warrant their attention and they began making liaisons with other disciplines. Thus the philosophy instructors involved themselves with people in the health and business fields and developed courses in Medical Ethics, Business Ethics, Logic for Computers, and similar cross-disciplinary activities. The literature instructors, acting in cooperation with instructors from other disciplines in the humanities, developed interdisciplinary courses combining literature with history, philosophy, art appreciation, and religion. They demonstrated the value of those courses, made them required for all degree candidates, ensured that students would write within the contexts of those courses, and built them into the fabric of the institution. In one 13,000-student community college in Florida, the humanities staff teaches sixty sections a year of an interdisciplinary humanities course. The course is required for students in all programs, including the occupational certificate areas.

Some mathematics instructors have taken similar steps. Davis (1980) studied the level of cooperation between mathematics and occupational technical faculty in designing and presenting mathematics courses for students in occupational programs. He found that these cooperative relationships were more likely to be developed where there was a high quality of informal communication among the staff, a process for development and review of content in the courses, and where the mathematics department took responsibility for initiating such course liaisons.

FUTURE

Mathematics instruction in the community colleges for at least the next several years will involve remedying the defects in mathematics preparation revealed by students graduat from high school. If recent curriculum reforms in the secondar schools have the desired effect, by the end of the decade the press for remedial studies in community colleges may have been lightened. At that time the attention of the mathematics curriculum planners will of necessity turn increasingly to ways of merging the study of mathematics so that the same courses become suited both for the liberally educated person and for the person intending upon immediate employment. This is essential because most students coming to community colleges want to be prepared for immediate employment but at the same time, they do not want to forgo their options for continued study. Furthermore, a person is both a worker and a citizen of the community and needs preparation in applied mathematics and in the broader concepts of mathematics that help the person as a citizen to understand developments in science and technolo outside the career field itself. And the career programs themselves have increasingly become feeders to the senior institutions. Many of the credits earned in the two-year colle occupational programs are acceptable for transfer.

Any such curricular reformation should take place in the community colleges themselves. There is no external agency organized for the purpose of revising collegiate studies in a manner that would better fit the colleges. The practitioners who have organized the mathematics laboratories, who are familiar with community college students and the internal politics of curricular reform, will have to undertake these changes in the direction of mathematics courses sufficiently broad and concentrated to satisfy the peculiarities of the institution in which they work.

Learning mathematics depends on an ability to imagine the future. Why manipulate those apparently sterile symbols? The student must appreciate the power of those symbols to effect technology, invention, the advance of knowledge in the sciences, the quality of life itself. This suggests a set of modules to be presented in conjunction with all formal mathematics classes and as part of the learning process in the mathematics laboratories. These short segments would be directed toward moving students away from a sole concentration on the symbolic language itself, toward understanding the power of that language. Basic mathematics is as much a service course to students in all programs as is basic English usage. The professionalized community college instructors would ensure that it was so presented.

REFERENCES

Astin, A.W. and Others, *Minorities in Higher Education.* San Francisco: Jossey-Bass, Inc., 1982.

Baley, J.D. "What We've Learned in Nine Years of Running a Learning Center." Paper presented at the Conference on Remedial and Developmental Mathematics in College: Issue and Innovations. New York, April 9-11, 1981. 13 pp. (ED 201 371)

Beckwith, M.M. *Science Education in Two-Year Colleges: Mathematics.* Los Angeles: Center for the Study of Community Colleges and ERIC Clearinghouse for Junior Colleges, 1980. (ED 187 386)

Brunson, P.W. "A Classroom Experiment Involving Basic Mathema and Women." *Two-Year College Mathematics Journal*, 1983, 14 (4), 318-321. (EJ 285 894)

Chang, P.T. "College Developmental Mathematics--A National Survey." Paper presented at the Annual Convention of the Mathematical Association of America, Charleston, S.C., April 15-16, 1983. 22 pp. (ED 234 841)

Cohen, A.M. and Brawer, F.B. *The American Community College.* San Francisco: Jossey-Bass, Inc., 1982.

Crepin, D.M. "A Developmental Mathematics Program for Communi College Students." Unpublished paper, Lower Columbia Col (Wa), 1981. 13 pp. (ED 210 076)

Davis, R.M. "The Development and Delivery of Mathematics Serv Courses in Two-Year Colleges." Unpublished Dissertation, University of Maryland, 1980. (ED 210 046)

Dyer, P.A. "American Mathematical Association of Two-Year Colleges Developmental Mathematics Curriculum Committee: Annual Report." Unpublished paper, American Mathematical Association of Two-Year Colleges, 1981. (ED 208 924)

Emerson, S. "RSVP Basic Math Lab: MAT 1992." Unpublished paper, Miami-Dade Community College, 1978. (ED 188 649)

Fast, C. "Making the Best Better: The LCC Individualized Math Story." Paper presented at the Oregon Developmental Educational Association. Eugene, Oregon, May 1980, 10 pp. (ED 187 371)

Friedlander, J. "Student Participation and Success in Community College Science Education Program." Paper presented at the Annual Meeting of the American Educational Research Association, Los Angeles, Ca., April 13-17, 1981. 15 pp. (ED 201 374)

Goldberg, D. "Integrating Writing into the Mathematics Curriculum." *Two-Year College Mathematics Journal*, 1983, 14 (5), 421-424. (ED 288 670)

Habib, B. "A Multi-Purpose Math Lab: A Place for All Seasons." Paper presented at the Conference on Remedial and Developmental Mathematics in College: Issues and Innovations. New York, April 9-11, 1981. 10 pp. (ED 201 361)

Mitchell, M.L. "Individualized Mathematics Instruction." In F.B. Brawer (Ed.), *New Directions for Community Colleges: Teaching the Sciences*, no. 31. San Francisco: Jossey-Bass, Inc., 1980.

Palow, W.P. "Technology in Teaching Mathematics: A Computer Managed, Multi-media Mathematics Learning Center." Paper presented at the Annual Meeting of the National Council of Teachers of Mathematics. Boston, Ma., April 1979. (ED 184 609)

Riley, M. "The Community College General Academic Assessment: Los Angeles District, 1983." Unpublished report. Los Angeles: Center for the Study of Community Colleges, 1984.

Rotman, J.W. "Developmental Mathematics and the Lansing Community College Math Lab." Unpublished paper, Lansing Community College (Mi), 1982. (ED 224 542)

Statements on Preparation in English and Mathematics: Competencies Expected of Entering Freshmen and Remedial and Baccalaureate-Level Work. Sacramento: Academic Senate for California Community Colleges, 1982. (ED 222 235)

Wood, J.P. "Mathematics Placement Testing." In F.B. Brawer (Ed.), *New Directions for Community Colleges: Teaching the Sciences*, no. 31. San Francisco: Jossey-Bass, Inc., 1980.

Yawin, R.A. "Remedial and Developmental Mathematics at Springfield Technical Community College's Mathematics Center." Unpublished paper, Springfield Technical Community College. Springfield, (Ma), 1981. (ED 213 456)

(DISCUSSION BEGINS ON P. 54.)

LET'S KEEP THE "COLLEGE" IN OUR COMMUNITY COLLEGES: MATHEMATICS FOR COLLEGE TRANSFER

by

Larry A. Curnutt

SUMMARY

LET'S KEEP THE COLLEGE IN OUR COMMUNITY COLLEGES
(Mathematics for College Transfer)

Larry Curnutt

Preparing students for transfer to four-year schools remains a significant part of the mission of most community college mathematicians. Freshman and sophomore mathematics offerings can be improved by including topics from discrete mathematics; this ought to be accomplished through revision and incorporation rather than wholesale change. Students of all majors can benefit from mathematics at the college level, and appropriate "liberal arts math" courses should be available for them. For the sake of mathematics and our students we should pay more attention to communication skills in mathematics classes. Continuing education of faculty may be the most important ingredient in effecting meaningful curriculum revision.

LET'S KEEP THE "COLLEGE" IN OUR COMMUNITY COLLEGES (Mathematics for College Transfer)

Larry Curnutt

More than twenty years have passed since I first stepped onto the campus of my hometown's community college. (They were called "junior" colleges in those days.) At the time, even though all of my rather vague career goals required baccalaureate degrees, I chose to attend a community college for financial reasons. Many of my classmates were in the same boat. Few of us had the luxury of worrying about the quality of the education we sought.

ENTRY LEVEL COLLEGE MATHEMATICS: AUDIENCE AND DEFINITION

Today's community college students comprise a much larger and more heterogeneous lot, and their reasons for attending a community college are often much more imaginative. Yet, despite the rich variety of two-year vocational programs available to them, the overwhelming majority of students who enroll in my community college mathematics courses (even at the elementary and intermediate algebra levels) still declare that transferring to a four-year school is their immediate objective. Fortunately, they study from better textbooks today, and, in general, their instructors are more extensively educated than they were in 1962. Staffed by educators who have consciously chosen to concentrate on the first two years of college (as opposed to "graduating" from the high school ranks or "dropping-out" of the research arena), community colleges are better equipped than ever to prepare university-bound students.

Exactly what constitutes appropriate preparation for college level work in mathematics, of course, depends on our definition of "college level mathematics." For some thirty years this issue has remained essentially static, for calculus has been synonymous with entry level college mathematics. (Elementary functions and finite math courses may have been available and even earned credit in the universities, but the attention paid to them and the status accorded them has been nominal in comparison to calculus.) However, the last three or four years have seen encroachment from two directions on calculus' predominance. First, most universities now admit that even some of their mathematics majors need to postpone calculus until their sophomore years [1], while users of mathematics require more and more access to all pre-calculus courses. Secondly, as computers have grown more accessible and computing more generally important, discrete mathematics has begun to find its natural (?) place in the curriculum. Some mathematician-educators (eg. Tucker and Lucas in [6]) have argued that a yearlong course in discrete mathematics should replace calculus for many students, majors and users alike. Since mathematization in many nontraditional disciplines seems inexorable, growth in the demand for post-high school precalculus mathematics in the lower division curriculum is inevitable (though I doubt that it will happen as dramatically as Tucker and Lucas propose). Thus, any reasonable definition of "college level mathematics" ought to include college algebra/elementary functions and finite mathematics, as well as calculus, introductory linear algebra and differential equations.

IMPROVEMENTS IN THE FIRST TWO YEARS OF COLLEGE MATHEMATICS

How should we in the community colleges meet this expanding, broadening audience in our courses? Our credibility with the universities hinges on the answer. A pertinent and unifying tenet bears repeating here: the thrust of freshman and sophomore mathematics courses should be to apply theorems rather than to prove them. (Yes, community college mathematicians have to exercise willful restraint in this area.) In deciding whether to include a particular piece of mathematics in a course, all of us who teach freshman/sophomore level mathematics should ask ourselves, "Is there a quick payoff for the students?" Usually, answers such as "Yes, in quantum mechanics." or "Sure, this will allow us to state the Fundamental Theorem of Algebra completely." must be discounted, respectively, as too distant and too incestuous to be compelling to our students. For example, do you teach DeMoivre's Theorem in your precalculus courses? How much time do you devote to complex numbers in your sequence of intermediate algebra, college algebra and elementary functions courses? More than enough, I suspect; they're in all the textbooks. At the risk of me learning something, I implore you to present just one genuine application, meaningful to lower division students, in which it is essential or illuminating to know that the roots of $x^2 + bx + c = 0$ are $r_1 \pm r_2 i$ rather than just that the roots are not real. I do not advocate that we banish complex numbers (some of my favorite eigenvalues are complex!), only that here is a place where we might buy some time for topics that are more relevant to the mathematics users in our classes who will not pursue mathematics beyond the calculus level. At a slightly higher level the same simple test can be applied to the formal definition of "limit" with similar results. Surely, no one will question the wealth of interesting mathematics and applications that can be managed with a sound heuristic understanding of the limit concept. Pedagogically, slogging through a few arguments will not increase the chances that limits will be used correctly or effectively. (Toward this end, consult [4] for a productive alternative.) Be sure, this is not a plea for training better polynomial differentiators in the community colleges; we cannot afford that image.

Although these examples suggest that there is plenty of room for improvement, I believe that the structure and implementation of the community college mathematics curriculum is basically solid. And why shouldn't it be? We've been practicing it for a long time; we've gotten good at it; and it does what it's designed to do -- to prepare students who are competent to do as well in schools of engineering, business, etc. as their university-trained classmates. However, community colleges have done no better job promoting discrete mathematics than most of the rest of the academic community. Perhaps we should all subscribe to an "affirmative action" plan that favors hiring mathematics teachers with specific interest and training in graph theory, combinatorics and numerical analysis. This plan should apply to mathematics, as well as mathematicians, so that when we do gain some slack in our courses, discrete topics will receive preference over additional continuous topics. Although there appear to me to be some inherent pedagogical disadvantages to a unified discrete mathematics course at the calculus level,* our existing courses could only be enriched by the inclusion of discrete methods. In many cases the machinery is already in place. Here are a few simple ways in which we could accommodate at least the spirit of discrete mathematics.

1. Use powerful, pervasive tools such as the computer, the Pigeon Hole Principle and systematic case exhaustion wherever possible and appropriate.
2. Return simple counting and probability to college algebra courses from where they disappeared in the not too distant past.
3. When the function concept is introduced in algebra, illustrate that domains don't have to be intervals. Use real examples, not contrived ones.
4. Don't skip the linear programming section in the precalculus book. Even though the Simplex Algorithm can't be fully explained here, require students to set-up linear programming problems and use canned software to crunch the numbers.
5. Numericalize (once in a while anyway). De-emphasize the classical theory of polynomial equations in favor of elementary approximate root searches. Emphasize setting-up Riemann Sums. Pay attention to error estimates.

* Not many students study mathematics because they really want to. Most students take a maximum of one year of mathematics beyond preliminary algebra. (There are some obvious exceptions, of course.) We do these students a disservice, if we totally ignore mathematics' discrete component throughout this year (just as we would, if we totally ignored its continuous component). Nevertheless, calculus enjoys features that make it an almost perfect mathematical vehicle for students at this level, and, thus, difficult for mathematics educators to let go of. (1) Calculus is intrinsically visual and dynamic. (2) The limit concept is a singularly compelling unifier. Does discrete mathematics possess aspects powerful enough to rival these and support a yearlong course at the freshman/sophomore level? My concern is that no matter how interesting and pertinent discrete problems are or how easy they are to state, counting, in all its conceptual simplicity, is still distractingly hard. I'm afraid that of the three principal aspects of combinatorial reasoning (systematic analysis of possibilities, exploration of logical structure of a problem, and ingenuity) that Alan Tucker lists in his preface to [9], the third one, INGENUITY, will weigh far more heavily on students at this level than the first two, which are more coachable. And it is not clear what we can do beyond the intermediate algebra level to prepare students specifically for such a course.

6. Pay sequences (including recursively defined ones) their just due in calculus. (If you must, here's a better place to worry about convergence proofs.)
7. Spend less time on special techniques for finding closed-form solutions to first order differential equations, and gain an opportunity to investigate finite difference equations.
8. Be sure to use Markov chains as an example in linear algebra.
9. Continue to provide a finite mathematics alternative for appropriate majors.

MATHEMATICS AND LIBERAL ARTS

Facility with intermediate algebra is not an unreasonable thing to expect of all bachelor's degree aspirants. Beyond that, even students whose majors hold no apparent applications of mathematics should be exposed to some college level mathematics. Why? Because it's good for them. Some of us still believe, despite the trend toward specialization, that those who study a little mathematics (along with the other liberal arts) are somehow (perhaps intangibly) better off than if they had not studied mathematics at all. (See [3] for a clear and amusing statement of an opposite point of view.) Specifically, requirements for this audience ought to include statistics and either a mathematics appreciation course [2] or one of the conventional college level courses (college algebra, finite math, calculus). Unfortunately, appreciation-type mathematics courses have often attained popularity as nonchallenging alternatives to greater evils. These courses have underestimated their audiences by concentrating on mathematical fun-and-games or consumer mathematics; or they have completely missed their marks by attempting deadly snatches at abstract structures. The real heart of mathematics appreciation courses should be a few "grabber" ideas, preferably in the form of applications/problems [8], around which two-week units could be built. The mathematical potential of foci such as exponential growth and decay and Leontief input-output analysis is obvious; and numerous excellent UMAP Modules are tailor-made for this forum.

Certain items do not appear explicitly in any mathematics curriculum, yet their effect on how well our students learn is so fundamental that they warrant broad attention in mathematics courses.

For example, the making of lists and tables of data may seem simple and artless to us, but actually requires nurturing. Organization of data/information, as an incipient part of problem solving strategy [7], hardly needs elaboration here, but it does deserve magnification in our introductory classes. The ability to organize data skillfully can also be a significant ingredient in clear communication, which (I'm slightly reluctant to admit) is not the province of English and speech departments alone [5]. Mathematics' part in a "communication across the curriculum" effort need not be overly ambitious. If we only succeed in convincing our students that scientists believe lucid writing and speaking are important and that we are serious about promoting them, then we will have taken a step in the right direction. Incorporating brief writing exercises in tests and assignments should not prove too painful; and insisting that students compose their classroom questions precisely in complete sentences should require just patience and a little direction from us. Understand that mathematics must remain the guiding objective. Therefore, any contribution to the communications enterprise will necessarily be a selfish one, for careless use of language can only hinder productive thinking (about mathematics). Furthermore, we're all well aware of how much an explainer learns in trying to clarify an idea or problem for someone else. Why don't we exploit this phenomena?

CONTINUING EDUCATION

Maintenance of established quality and effective curriculum revision must come from an informed faculty, not imposed upon them. Community college administrations have not been notoriously supportive of faculty professional development except on a limited local scale. (Some have even been accused of fostering anit-intellectual environments.) Therefore, community college mathematicians need to make special efforts to keep up-to-date by reading journals, solving problems, attending professional meetings and enrolling in classes. Our national mathematics organizations must continue to recognize their two-year college constituency by sponsoring appropriate minicourses, publications, etc. Administrations must be convinced of the benefits to their institutions of funding rich sabbatical leave programs, faculty travel to national conferences, personnel exchanges with industry and unlimited faculty access to new technologies such as microcomputers, word processors, video discs, etc. In the absence of these efforts changes in the curriculum will be cosmetic at best; with them it will evolve naturally and meaningfully.

REFERENCES

1. CUPM. Recommendations For a General Mathematical Sciences Program. The Mathematical Association of America, 1981.

2. CUPM. "Minimal Mathematical Competencies for College Graduates." The American Mathematical Monthly. 89(1982), pages 266-272.

3. Underwood Dudley. "Mathematics: Who Needs It?" San Francisco Chronicle. April 4, 1984, page 34.

4. Ross Finney, Dale Hoffman, Judah Schwartz, Carroll Wilde. The Calculus Toolkit (software). Addison-Wesley, 1984.

5. Dorothy Goldberg. "Integrating Writing into the Mathematics Curriculum." The Two-Year College Mathematics Journal. 14(1983), pages 421-424.

6. Lynn Arthur Steen, ed. Mathematics Tomorrow. Springer-Verlag, 1981.

7. G. Polya. How To Solve It (2nd ed.). Princeton University Press, 1957.

8. Dalton Tarwater, ed. The Bicentennial Tribute to American Mathematics. The Mathematical Association of America, 1977.

9. Alan Tucker. Applied Combinatorics. John Wiley & Sons, 1980.

(DISCUSSION BEGINS ON P. 64.)

A NEW START FOR MATHEMATICS CURRICULUM

by

Alan Tucker

A NEW START FOR MATHEMATICS CURRICULUM

by Alan Tucker
State University of New York at Stony Brook
Stony Brook, New York 11794

SUMMARY

This writer advances the thesis that a major re-thinking of the mathematics curriculum is needed at secondary school and upper-division college levels. Two-year colleges suffer the most from the problems in the current curriculum and may be the best place to initiate change.

I. IT IS TIME FOR SOME CHANGES

Two of the dominant issues in mathematics education today are, first, curriculum and student performance in secondary school (7-12) mathematics, and second, the role of computers in mathematics education. This writer believes that the curriculum can be improved by the use of computers and the mathematics related to computers.

The academic community and employers are not complaining about collegiate mathematics curriculum or the mathematical skills of college graduates. Rather most problems facing colleges, two-year and four-year, seem to be the result of problems in the secondary schools. The lack of basic skills in high-school graduates especially affects two-year colleges, where over 40% of mathematics enrollments are at the level of intermediate algebra or below. These remedial courses are often a trying experience for both students and faculty because the pace is too fast (remedial courses tend to have a lot of ground to cover, but if students did not learn the material when taught at a normal pace, teaching it faster is not going to help) and the students are unmotivated (the courses are a general studies requirement forced on the students).

A basic conclusion of the recent CBMS Conference Report, New Goals for Mathematical Sciences Education, was "the current organization of topics, courses, and course content needs to be reformulated for all students [for pre-college mathematics]". In particular, this writer suggests that the pre-calculus orientation of the high school mathematics, and hence of remedial collegiate mathematics, may now be all wrong. Little more than the basic concepts of calculus are needed by the average student who does take calculus, say, for advanced courses in an economics major. On the other hand, the mathematics related to computers and mathematics dependent on computers for computation (such as statistics and linear models) provide topics for study that are intellectually rigorous and much more widely applicable.

Calculus is something engineers and physical scientists use extensively. However, its fundamental purpose for many other students is now a benchmark course for measuring academic rigor in mathematics. It is required of all pre-meds but never used in medical school. "Calculus develops mathematical maturity" is the phrase heard so often. People are impressed when a college student comes home and tells his/her friends and parents that he/she is taking calculus. But the calculus course of today is much more superficial than the calculus course of thirty years ago, the course that gave calculus its good name.

Ralston and others have argued persuasively that calculus is not the only way to develop mathematical maturity (see the Proceedings of the 1983 Sloan Conference on the First Two Years of College Mathematics, published by Springer-Verlag). The discrete mathematics of computer science and other new applied mathematics has equally attractive maturity-building content. The case for changing from a calculus focus seems strongest for those who will never even take calculus, but are forced to take preparatory courses leading up to calculus.

Other college departments often encourage mathematics departments to keep calculus in its focal position, but these same departments are nibbling away important new areas of the mathematical sciences by teaching their own courses in statistics, computing, or mathematical modeling: psychology faculty may tell a Math. Dept.,"We will teach our own statistics courses for our average students who learn statistics better in a psychological context, but we will encourage our very best students to take calculus and a rigorour probability theory course from you." Many mathematicians find this vote of support for calculus very assuring. This writer does not view efforts to lock lower-division mathematics into pre-1900 topics very reassuring (while others teach the mathematical sciences of the 20th century).

II. TWO-YEAR COLLEGES SHOULD TAKE THE LEAD IN CHANGING MATHEMATICS CURRICULUM

The current interaction between colleges and secondary schools in trying to change mathematics curriculum is a classic example of negative feedback (with "negative" used here in the sense of encouraging each other to do something the wrong way). Secondary schools look to colleges for guidance in their curriculum, since they are supposed to prepare students for college courses. But the colleges now are looking to the secondary schools for their curriculum: remedial curriculum is based on high school courses. There are reasons to believe that secondary school mathematics and the remedial college curriculum are stuck in a rut and each is digging the hole a little deeper for the other (these reasons include the exclusive pre-calculus orientation of remedial courses; lack of computing or probability/statstics or applications; emphasis on routine but unmotivated, algebraic techniques, justified by "you will need this for subsequent college math courses").

The two-year colleges are even worse off, since they look both to four-year colleges and high schools for guidance. They are not expected to set standards or take any initiative about the training of students they admit from high schools or students they pass on to four-year colleges.

Two-year colleges should turn the tables around and take the initiative in developing new curriculum and teaching modes. Two-year college faculty have a much greater interest than the four-year college faculty (whose training was research-based) in the transition high school-college mathematics curriculum and its teaching. Most universities handle lower-division classes in huge lectures and/or TA-run sections, where innovative teaching is impossible. TYC faculty also have many advantages over high school teachers, in their training, the classroom environment, and the absence of standardized tests such as College Board exams (which stifle curriculum innovation). The problem of poor motivation found in two-year college students is even worse in high schools students.

Two-year college students are typically older and more mature than four-year college students. Most are in a two-year college for a purpose, not forced to be there by law or parents, although they may be forced to take a particular math course as a requirement of their major or as a general studies requirement. The mathematics courses for these students should respect their maturity and motivation. They should have clearly-explained, meaningful goals. There are so many jobs in society today that depend on quantitative reasoning that it should be easy to design mathematics courses that develop useful (marketable) skills and/or modes of reasoning and challenge students' maturity.

Thus, two-year colleges should take the initiative in mathematics curriculum changes that in turn will change the high school curriculum. High schools certainly cannot be expected to initiate change. They worry about training students for the College Board exams that are written by college people. Some four-year colleges are starting to make changes, but the colleges involved are the most selective colleges and their students are so different from the typical high-school graduate that their curriculum ideas may take years to reach two-year colleges and high schools.

III. PRINCIPLES FOR A NEW CURRICULUM

The question is what mathematics should the typical two-year college student know? Closely related is the question, what mathematics should the typical high-school graduate know? Instead of a watered down version of what a pre-engineering student needs, our answers should put the typical TYC student in center stage. We may give the pre-engineering student a separate curriculum or maybe an advanced version of what the typical student gets.

The first step in revising the mathematics curriculum is deciding what the fundamental objectives of the curriculum are. This decision brings us immediately to the inevitable trade-off between reasoning ability and problem-solving versus mathematical facts and technical skills. The NCTM _Agenda for the 1980's_ gave problem-solving its top priority. This writer agrees with that priority. But problem-solving takes time. Whatever material is taught, it will have to be taught more slowly to make time for problem-solving. Implicit in a problem-solving emphasis is that the problem-solving should involve relevant applied mathematics: statistics, systems analysis and other modeling, and most of all computing. One further tenet of the writer is that the mathematical concepts and techniques should be well motivated and meaningful to the student. This also takes additional time.

A curriculum of relevant mathematics will include new topics, but it will in large measure be current topics taught from a different point of view. There are many underlying mathematical skills needed in solving applied problems, in particular the symbolic reasoning and manipulation of algebra. But these skills should not be taught in isolation from their uses. Instead, some applications should come first. The usefulness of symbolic manipulation must be made clear. If a course just teaches students to genuinely appreciate the power of symbolic reasoning and manipulation, then that alone would be a major achievement.

There are two ways in which algebra is naturally motivated. One way is through computer programming. Programming the solution to a set of numerical problems is mostly algebra. The other way is by trying to do problems without using algebra. Let students work problems that can be done "the long way" by trial-and-error or tedious calculations, but which are quickly solved with algebra.

Students should see that algebra makes problems easier; that moving away from the specific numbers of a real-world application into a more abstract setting can be the best, fastest, and most insightful way to solve that practical problem. Of course, formulating one general solution for a class of problems is also what computer programming is all about.

An analogous situation is the way grade schoolers accept the need to learn multiplication tables. Initially they compute simple products like 3×4 by repeated addition. But long multiplication problems like 334×52 win students over to the advantages of memorizing products. In a similar way, students should be encouraged to experiment with "the long way" solutions, even using computers. For example, the exponential function e^x can be motivated by compound interest calculations with yearly, then monthly, then hourly, then second-by-second compounding; or plotting, in finer and finer time intervals, the size of a population of bacteria whose growth rate is equal to its current size. The abstract formulations should come when students see that they need them to solve problems. The obvious dividend of this approach is that students would leave these mathematics courses with a positive attitude towards mathematics and, more generally, abstract reasoning. They would be more accepting of mathematical approaches in other subjects and would be more willing to take more mathematics courses later in their careers.

Summarizing, the two-year curriculum should emphasize:

* Problem Solving
* Meaningful Mathematical Topics
* Practical Applications
* Proper Motivation of the Mathematics

IV CONTENT OF NEW CURRICULUM

As noted in the previous section, algebra must continue to play the central role in the high school curriculum. Algebra should be taught in conjunction with computer programming and it should be used to organize and simplify the solution of applied problems. To compute a table of values for net profit of a company with various possible sales rates and production costs (as in a "spreadsheet program"), a set of algebraic equations must be formulated to describe total income, total expenses, etc.; and then these equations must be programmed.

Probability/statistics and logic are two other basic topics, which also should be taught with a heavily applied emphasis. For example, propositional logic can be embedded in many programming questions: what combinations of circumstances will cause the program to come to a certain statement? Probability and statistics have limitless applicability, and an empirical approach based on applications should be used.

Finally, the curriculum should include matrices and linear models. This writer's interest in this topic is based on a very deep philosophical theme. In the 19th century, calculus was the model of what a scientific theory should be. All other scientific theories were expected to have the simple, elegant structure and predictive power of calculus. Today, we know that this is a very naive view. Now we expect that most interesting phenomena are inherently complex and have many input variables and many output variables. Think of the molecular biology of a cell or of a DNA molecule; think of the design of a spacecraft; think of an airline's computerized reservation system.

Matrices are the simplest mathematical model for such systems of organized complexity. A linear model that hypothesizes linear relationships between a set of inputs and a set of outputs (in the form of a matrix of input/output coefficients) is easy to build for almost any complex system, and its behavior is easy to simulate on a computer. A matrix of input/output coefficients is the first-order approximation of any activity that can be measured numerically. This linear-systems way of thinking about activities around us is the basic quantitative philosophy of the modern world, and as such, should have an important role in the curriculum.

These topics will develop students' reasoning abilities in a way that they can respect, and that will lead to a general college math requirement that can be considered just as essential as a general college English (writing) requirement.

V EXAMPLES OF NEW TOPICS IN THE MATHEMATICS CURRICULUM

* As noted above, algebra would be introduced in connection with computer programs to do numerical computation, and logic could be taught in connection with logical possibilities for program flow; logic and Boolean algebra are also used heavily in finite probability and combinatorial analysis.

* One of most common algebra examples of two equations in two unknowns involves the speed of a canoe going upstream and downstream. The new canonical example could involve two companies A and B with current sales of s_A and s_B with $s_A > s_B$ and current growth rates (for sales) of g_A and g_B, respectively-- specific values would be given to s_A, s_B, g_A, g_B. We want to know when (if ever) B's sales will surpass A's sales.

* Geometry should arise in the context of computer graphics. Drawing shapes and moving them around a screen is almost wholly analytic geometry. Euclidean geometry would have a minor role, in secondary schools, in this curriculum.

* Logarithms and exponential functions would be taught in connection with exponential growth models and binary-tree data structures (for dictionary searches).

* Probability and statistics should have an exploratory spirit, using statistical computer packages, to look for trends in data.

* The role of trigonometric functions needs to be re-thought. Trig is used almost exclusively by students in engineering or physical sciences, and there in connection with Fourier series and complex-valued solutions to differential equations. Perhaps Fourier series should be introduced early in calculus, and symbolic integration programs used to compute Fourier coefficients. Then one could plot (on a computer) approximations to a function f(x) using the first k terms of the Fourier series for f(x). Elementary time series

analysis might be attempted.

VI Concluding Remarks

The suggestions here for a new curriculum are fairly radical. No texts now exist. As was done with four-year colleges, the Sloan Foundation and National Science Foundation should support the efforts of a few two-year colleges to develop new mathematics curricula. The two-year colleges must obtain the backing and professional cooperation of the principal four-year institutions to which their graduates transfer for continued college study.

REFERENCES

1. "The Mathemaical Sciences Curriculum K-12: What is Still Fundamental and What is Not", Report of Conference Board of the Mathematical Sciences to the National Science Board Commission on Precollege Education in Mathematics, Science, and Technology, December 1982.

2. "New Goals For Mathematical Sciences Education", Report of Conference sponsored by the Conference Board of the Mathematical Sciences, November 1983.

(DISCUSSION BEGINS ON P. 70.)

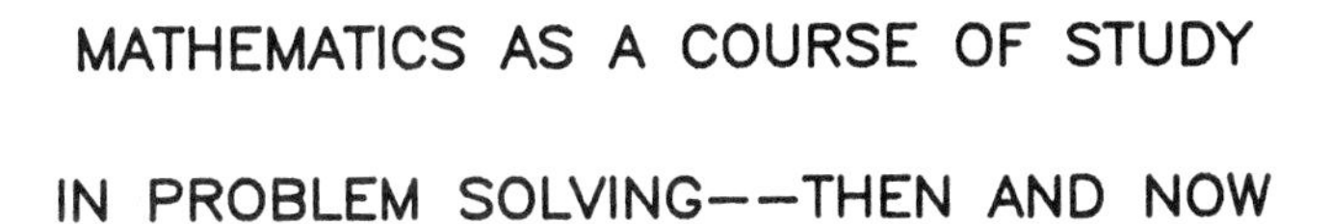

MATHEMATICS AS A COURSE OF STUDY IN PROBLEM SOLVING--THEN AND NOW

by

Wade Ellis, Jr.

SUMMARY

The mathematics curriculum in the first two years of college is a tool created to assist in solving problems. The enormous computational capability provided by digital computers has increased the number of problems that can be addressed and solved using elementary mathematical concepts to include problems from the social sciences, business, and the art and science of computing. The current curriculum has not yet adjusted to the larger variety of problems that fall within the scope of mathematics that might be taught in the first two years of college.

The author compares the content and style of the mathematics curriculum at the end of World War II with its content and style now. The impact of the development of the computer on the mathematics curriculum in the first two years of college is discussed. Directions for curriculum change are suggested, and the implications of curriculum change for teacher currency, concept presentation, and curriculum content are explored.

Mathematics as a Course of Study in Problem Solving: Then and Now

Wade Ellis, Jr.
Mathematics Department
West Valley College

Mathematics has has been a part of the college curriculum for a long, long time. Historically, from the Greeks through modern times, some knowledge of mathematics has been part of every educated person's experience. In more recent times, the usefulness of mathematics as a problem solving tool has been emphasized by the rise in technology beginning with the Industrial Revolution. This paper will focus upon this problem solving aspect of mathematics -- as it might have been at the end of World War II and as it is now.

Then

In the mid-1940's, the high-school curriculum included, with some variation, the following:

Algebra I, Algebra II, and Geometry

Trigonometry and Advanced (College) Algebra

Solid Geometry

The first three year-long courses were intended as an introduction to the symbols and pictures used in mathematics. The second two courses were intended as the finishing touches for students intent upon an engineering, science or mathematics career in college. Great importance was placed upon speed and accuracy in the manipulation of symbols and upon geometric proof using pictures. The able student finished this curriculum well prepared for the college mathematics curriculum. Non-science students required no further mathematical training in college beyond the first three years of high school.

The beginning college curriculum was tied to the engineering and science curriculum and was an integral part of it. The following courses comprised that curriculum:

Calculus of one variable
Calculus of several variables
Differential Equations

These courses were intimately tied to the engineering curriculum in that problems to be solved in the engineering and physics textbooks used techniques taught in these courses. The ideas that these textbook problems were contrived to give the student some success in the problem-solving process was well-known, encouraged and applauded. In the best of circumstances, students became confident of their problem solving ability and the mathematical resources that were available to them. In less auspicious circumstances, mathematics was seen as a burden that students endured until they were allowed to use the formulas that appeared in the CRC Handbook and other scientific and engineering handbooks. If you could plug things in, then you would do all right.

Problem solving based on this curriculum was apparently successful as the group of students educated under this course of study was able to put a man on the moon, no small problem-solving feat. Calculus and differential equations, the mathematics of Newtonian physics, were the keys to the solution of that problem. The combined mathematics-physics-engineering-science curriculum was a raging success.

But times change and so do the problems needing solutions. For example, are the current problems still best approached from the point of view of Newtonian physics, or are other methods more fruitful? We no longer teach the Ptolemaic system of planetary motion because improved observations showed Ptolemy's explanation to be inadequate. Later, again with planetary motion (Mercury), Newtonian mechanics was shown to be inadequate. Newtonian physics is still a marvelous creation and is correct (and will remain so) in an enormous number of problem settings. But there are other mathematical notions that are needed to deal with physics and for the illumination of the underlying concepts of Newtonian physics the rigors of calculus may no longer be required. The differential calculus was needed to allow symbolic calculations to be substituted for numerical ones. Now, the digital computer with its enormous numerical calculation ability might be a better tool for introducing Newtonian mechanics. Quantum mechanics, which is now taught in the engineering physics courses at community colleges, requires more than calculus.

With shifts in the physics and engineering curriculum, must we still require that students learn the intimate details of differential and integral calculus and not learn and benefit from the numerical techniques that are now feasible with digital computers? Are the years of Algebra, Geometry, Trigonometry, Calculus, and Differential Equations really required or is there a short cut or a short circuit for our students?

Now

The current curriculum has changed little from the post-war years. The names of the courses are a little different and Solid Geometry has disappeared as a name and as a topic in the high-school curriculum. In colleges, a watered-down version of Linear Algebra is now taught to some sophomores. Basically, we are still teaching the same general topics to students now that we did in 1945.

The details have changed nonetheless. The emphasis on epsilon-delta proofs has increased enormously. The symbol epsilon does not appear in the 1941 revised edition of *Elements of the Differential and Integral Calculus* by Granville, Smith and Longley, and it has only two sections occupying 5 pages on the limit concept. In Leithold's 1981 fourth edition of *The Calculus with Analytic Geometry*, there is a 77 page chapter on the limit concept. Though the emphasis on epsilon-delta proofs is currently decreasing, it is far greater than it was in 1945. Function notation is more widely used than in the past in mathematics courses (but not in engineering/science courses). The quality of textbooks has improved immensely with more numerous and better examples and graphs, though we still teach mostly by lecturing with little dependence on a textbook except for problems. Unfortunately, the topics are still the same.

Have the problems we wish our students to solve changed? Oh my, yes! The business curriculum and the social science curriculum are now bristling with problems that need mathematical tools for their solutions. The exploding computer science curriculum is placing enormous demands upon the mathematical problem solving capability of students we teach.

The business curriculum appears to require strong data analysis problem-solving methods including graphically displayed data analysis techniques. This places a premium on understanding concepts of data analysis and diminishes the need for strong manipulative skills in algebra. Statistics is important and so are the ideas of increasing and decreasing functions, but from a graphical point of view. Linear programming is important as well but the analysis of the results of the calculations is much more important than the calculations themselves. The same needs are faced in the social science curriculum.

The computer science curriculum, on the other hand, places much more emphasis on manipulative skills in dealing with finite state machines, function notation and proofs of an inductive nature. In addition, new topics such as semigroups, graph theory, algorithms, and combinatorics are part of the curriculum. In addition, a formal programming language is needed to concisely represent and understand algorithms.

As an example of the new pressure placed upon the mathematics curriculum, let us consider some topics from the table of contents of a recent text on mathematical structures for computer science. We find topics such as:

Induction and Recursion

Relations and Digraphs

Connectivity and Warshall's Algorithm

Posets

Representations of Special Grammars

Undirected Trees

Products and Quotients of Groups

Finite-State Machines

Coding of Binary Information and Error Detection

These topics are of special interest to us because they are from a sophomore textbook, written by mathematicians, to be taught by mathematicians. It is not a list of topics from an isolated lecture-note type presentation of an experimental nature, but rather a course that has been taught for several years at Drexel University. In fact, the Computer and Information Science Department at my community college is asking the Mathematics Department to teach a course like it this coming year.

The Impact of Computers

The mathematics community has a problem. We have an extremely well-thought-out and tested curriculum in many versions (The Ohio State University teaches 9 different varieties of Beginning Calculus). It does not satisfy the changing needs of perhaps 75% of the students we teach although it is still a marvelous course of study for students of engineering, physics, science, and mathematics. The computer has in large part brought on this problem in two ways. First, the availability of fast and accurate computational devices has increased the accessibility of problems whose solutions use established mathematical techniques but require massive computational capability (if done by hand or with a mechanical desktop calculator). The use of computers to solve such problems has increased the effectiveness of investigations in the social science and business disciplines and thereby changed the curriculum needs of the students in these disciplines. Second, the complexity and development of computers and computing has required the application of mathematical ideas not contained in the "old" mathematics curriculum to problems in computer science. In some cases, computer science problems have required the creation of new mathematical ideas for their solution. These new ideas will become candidates for inclusion in the mathematics curriculum.

Microcomputers have also given us a new tool to use in presenting information and ideas to students. The availability of a microcomputer (with a suitably large screen display) can greatly enhance the presentation of material in class through rapid calculation and accurate graphs. The microcomputer can also be used to present information in a highly structured manner to individual students in laboratory settings. Microcomputers can also provide exercises in discovery learning of mathematical concepts.

Finally, the availability of symbolic manipulation programs on microcomputers reduces the need for computational skill with symbols in much the same way the hand-calculator reduced the need for computational skill development in arithmetic. This new availability calls into question much of the course of study in high schools and colleges and is both a blessing and a

curse. Textbook writers and manufacturers have a vested interest in the current curriculum. In our bottom line society, they are trend setters. It will be difficult to change their minds about what is to be in the curriculum. We, as teachers, also have a vested interest in this curriculum since we have learned over many years how to deal effectively with it: how to present it; how to measure student success in it; and how to remediate student difficulties with it. So much for the curse. The blessing is that the reduction in the need for computational skill development throughtout the curriculum provides us with the needed time
to introduce the new topics and approaches that will be required if we serve our students well.

The Computer in Matheamtics

For those of us who believe that computers like calculators are going to go away, let me say that mathematical research is driven in large part by examples, and examples often require calculations. The more complex these calculations can be, the more quickly patterns can be discerned and tested. Computers can help and are helping in doing these calculations, both numerically and symbolically. They are here to stay.

Teacher Preparation

The content, presentation styles, and methods of mathematics are changing rapidly. As a teacher in the field, I feel woefully underprepared to teach the new topics with the newly available teaching techniques, but am extremely excited about the possibilities. I do not know enough, although I have taken advanced graduate level courses in combinatorics, real and complex analysis, number theory, and abstract algebra. I have no idea what Warshall's Algorithm is and no guess as to how to teach students about it.

I have spent an entire sabbatical year studying the uses of computers in teaching and have taught a course entitled "The Microcomputer as a

Research Tool in Mathematics." I have written textbooks about the computer language BASIC. Still, my hands begin to shake when I attempt to present mathematical ideas using a computer in a classroom. There is so much that can go wrong or right that I know nothing about. Students are frequently asking easy questions that I have no idea how to answer.

We can teach the current crop of novice teachers to know more of the content and we can teach them to use the current new techniques. But how much impact will these new teachers have? For that matter, when is the last time we saw someone who was embarking upon a career in teaching mathematics?

We are currently going through the hiring process for two positions at West Valley College. All of the applicants completed their training and coursework 3 to 20 years ago. Only one was completing his coursework this year and this applicant was a retrainee from another discipline. Many of the applicants had Ph.D.'s from other disciplines vaguely related to mathematics. The message, for me at least, is what we have is what we are going to have for the forseeable future. We have seen the enemy and it is us.

Because the curriculum is changing so rapidly, we must devise methods of encouraging the mature mathematics instructor to learn on the job. Since nobody, in my very humble opinion, knows where any of this is going, we must encourage ourselves to delve deeply into the new situation in mathematics each and every day. This means that we must seek reduced teaching loads so that we may have the time to learn the new material and experiment with the new techniques. In my opinion, workshops will not work because they give the false impression that there is the possibility of a quick fix for these problems. Each department must form itself into an effective learning environment for the new mathematics that most of us do not know. We must create an experimental learning environment with ample access in our offices and homes to the new computer technology so that we may internalize what it is we teach in the same way that we have internalized the limit processes of the calculus.

The Future

As the skills needed for problem solving change under the impact of computer computation (both numeric and symbolic), we must change the focus of ***mathematics as a course of study in problem solving*** from the manipulative skills required to solve differential equations to the conceptual skills needed to manipulate ideas. Thus, in the future, we must concentrate on modes of thinking and not on skills in manipulation. We should concentrate on six thinking skills:

Axiomatic Thinking (Theorem/Proof)
Programmatic Thinking (Large Problems into Small)
Algorithmic Thinking (Finite Solution Processes)
Probabilistic Thinking (Statistical Approaches)
Eclectic Thinking (Are there several ways to solve a problem?)
Constructive Thinking (Can a simulation or model be created?)

If we are to restructure the curriculum, then we need to consider those concepts and topics which introduce, develop and enhance these patterns of thought. The use of these mathematical modes of thought in problem solving is the direction in which I believe the first two years of college mathematics should go.

In addition, we should concentrate on preparing our students to read and understand mathematical ideas of all kinds and then to use these newly acquired ideas in solving problems. Along with this, we need to encourage our students to write and speak mathematics well so that as they acquire new knowledge they can pass it on to others. I hope in this endeavor we are able to encourage our students to spell it all out rather than use the terse and elegant journal style we all admire so much.

If we begin to change the curriculum in these directions, then we will find that the current lecture form of instruction will not suffice. We will need to break away from the teacher as all-knowing knowledge provider and begin to think of the teacher as guide and advisor in a student's quest for knowledge. We will begin also to depend more upon guided discovery learning techniques and written mathematical materials. Perhaps, here again we will call upon the computer to assist us in structuring learning environments that will gently push the students toward improving their powers as problem solvers using mathematical modes of thought.

The future is bright with challenges. Let us hope we can meet them.

References

Computing & Mathematics, Editied by James T. Fey et al., National Council of Teachers of Mathematics, 1984.

Discrete Mathematical Structures for Computer Science, Bernard Kolman and Robert C. Busby, Prentice-Hall, Inc., 1984.

The Future of College Mathematics, Edited by Anthony Ralston and Gail S. Young, Springer-Verlag, 1983.

Mathematical Structures of Computer Science, Judith L. Gersting, W.H. Freeman and Company, 1982.

New Directions in Applied Mathematics, Edited by Peter J. Hilton and Gail S. Young, Springer-Verlag, 1982.

(DISCUSSION BEGINS ON P. 76.)

DISCUSSION

A CASE FOR CURRICULUM CHANGE

1. Four-Year Institutions Ignore Two-Year Institutions at Their Own Peril
2. Liaisons with Secondary Schools
3. Students Know What They Know
4. Professionalism
5. Curriculum Reform
6. Is Failure to Transfer an Indictment?
7. Progress and Money, Money, Money
8. Math--The Heavy, Curriculum Disaster, and Problem Solving

FOUR-YEAR INSTITUTIONS IGNORE TWO-YEAR INSTITUTIONS AT THEIR OWN PERIL

Cohen: Community colleges enroll almost half of all people beginning college in this country. So people in other institutions ignore two-year colleges at their own peril. What happens in two-year colleges affects senior institutions. Now, just to say that is not of great moment because the senior college people ignored the secondary schools for generations at their own peril but without realizing it. For years people in graduate departments would say, "If you are good, we are going to train you as a scholar. If you are not good, go get a teaching credential, and go down to the lower schools." There is a mind set in graduate schools among many academic departments, and they pay the price because ill-prepared students come into the institution, and why are they unprepared? The answer is perhaps because the graduate faculty was sending their least well-prepared students to teach. So the community college is ignored at the senior institution's own peril.

Nearly all of the students coming into two-year colleges come from secondary schools. Less than ten percent of them transfer to a university. This is ignoring what should be a major connector point, and that is curriculum articulation between the secondary school and the community college. It is much more important to connect the curriculum between those two institutions than between the two-year college and the university.

Liaisons with Secondary Schools

Cohen: On a nation-wide basis, fifty percent of the mathematics enrollments in community colleges are in mathematics through algebra; that is, remedial arithmetic or beginning algebra and geometry. To learn how to teach that group well, to teach them what they did not learn in their first twelve years of school, you do not need to look to the university. The university does not have any bright ideas about that. Community colleges have to generate their own pedagogy and their own curriculum patterning, and they have to do it if they are to seek a liaison with any other institution. That liaison should be with the local secondary school because that is where the failure has occurred.

Students Know What They Know

Cohen: We built a testing instrument* for a different purpose. We wanted to know how much cohorts of students know. The purpose was to find out how much more math students know who have taken one math course than do students who have taken no math courses. If they have taken two courses, how much more do they know than students who have taken one course? If students are older, what is their math knowledge relative to younger students, and so forth. If they live in the East, how much more or less math do they know than students who live in the West? We generated cohorts. It was not an attempt to discriminate between individuals; it was an attempt to find what entire cohorts know. When we built the test that way, we used a different mode of item selection; if everybody knows an item, that's fine. We built a test of student knowledge in science, social science, math, English usage, and humanities. And we gave it this past year to a sample of some thousand students in four, large urban community college districts in Los Angeles, Chicago, Miami, and St. Louis. We got our math items from the

*See p. 10.

National Assessment of Education Progress, and we went through a great selection process trying to find items that were of the type that community college students might be expected to know. Among the more interesting results is the following: We gave them a question which said, "Compared with other students at this college, how would you rate your ability to solve problems using algebra?" The students who rated themselves excellent had a mean score of 6.17; the students who rated themselves poor had a mean score of 3.82. That is the greatest discrimination among cohorts in terms of their scores on the mathematics portion of this test--"How much math do you know?" And they know. They know what they know.

Our concern was that if students take a course, do they learn anything? I mean, how much more do students know who have had one course than students who have had no course? We were concerned with getting a measure of the value added or the ability gained as a result of taking community college courses. And the reason for that was the university, as we heard at this table, tends to ignore the community college. And when you wake them up, and they begin to think about the community college, the first response from a sizeable number of people who study universities and who speak from the university perspective is: "They do not learn anything there. Any students who are serious about their education would not go to a community college. They would go out of secondary school and go directly to a university." So we wanted to get some kind of measure of how well they learn something when they take a class. They know more than if they did not take a class, which is my definition of learning.

Rodi: Perhaps placement tests don't add a whole lot to what one might get by simply asking students how much algebra do they think they know, or how much math did they have in high school.

Case: But the validity of self-placement could not be deduced from this in any way unless you remove the three-fourths that were not entering students for the sampling.

Cohen: We did that; we rotated that and the beginning students are every bit as good as the ones that have been there. I do not want to carry on against placement procedures, but students know what they know.

I'm not saying, "dump your placement and just ask the students." I know better than that. But if you think about the way placement tests are generated and scored--they are just that. Students go to advanced, they go to introductory, they go to remedial. That is what they are used for--making discriminations between students. Now, why do they work so well? Because you take students and tests that are nicely curved from the population. You bring them into a class and say, "Now I'm going to deal with you and give you a test. You get an A, you get a B, and you get a C." That is, you are taking a normative test, and you are using it to predict a normative grade. What we are doing is to try to cut the world in a completely different way: "How much do you know as a cohort?" Not by, "Do you know more than he does?" This is a completely different approach. And I should say that in presenting these findings, this is the most difficult concept to explain. People in education do not understand how to measure student learning in an absolute fashion. The entire system is arrayed for purposes of discriminating between people.

Professionalism

Long: I found one of your comments most distressing. In your paper you said that the faculty opted for reduced teaching loads as the most important thing for their own professional well being. It seems to me that negotiating away sabbaticals and travel opportunities are exactly the wrong way to go if you are going to improve as a professional person. At the same time you go ahead and talk about professionalism in the sense of faculty becoming really professional teachers as opposed to the view of the university professors who are devoted to increasing their knowledge in the subject area. And you are suggesting that we can have highly professional teachers, and I guess I've thought about that and think it's so. I guess I don't see quite so much harmony in being skilled in the subject matter and in teaching. You better be skilled in both.

Cohen: In many of the reports that I've seen and many of the campuses that I have visited, it appears that there is a more highly professionalized form of instruction going on at the remedial/ math lab level than in many of the advanced classes from the standpoint of the discipline of instruction.

Long: I guess my bias is not that way. You have seen a certain professionalism in dealing with the remedial student, perhaps which is not so appropriate in dealing with more advanced students. So, I don't see that that necessarily says one is more professional than the other. But the thing I think that is terribly needed in community colleges is more professionalism among the cohort of community college faculty.

Rodi: You've got to remember that many community college faculty members, who are teaching fifteen hours a week with classes ranging from 35 to 40 students and quite frequently even larger, might be dealing with 200 different students in a semester and might have as many as three, and in some cases four preparations, and they don't have the assistance of graders. By comparison with university standards, these people are working extremely hard. There is something that influences their reaction about what is most critical, and that is to have a reduced teaching load. And in some cases it's their survival. They say, "I can't keep this up indefinitely as a human being." In other cases, they see the need for that profession-alism that you are talking about--reading, interacting, going to meetings--but they know they can't do that, or they can't do it in a truly productive way, unless they are given some release time from these heavy teaching loads. So, I think that if they are given choices: to get the number of students down, or the number of preparations/classes down, or have sabbatical leaves and time to attend professional meetings, they may opt for reduced travel time or no sabbaticals. They should not negotiate that away, but I think you've got to expect community college instructors to become more professional in the same sense you envision it and in the sense everybody here envisions it. You've got to give them adequate time to do it.

Long: It's not that rosy on the other hand in the four-year institutions. I normally teach two classes in a semester, and I may well be handling 250 students in the process. It takes a certain amount of grimness and a certain amount of dedication on the individuals' part--mine, yours, and everyone else's--in order to be professional. At the same time, I thoroughly agree with you that there needs to be more recognition on the part of the administration that time is needed for the faculty to be professional.

Fusaro: There is a danger here. Having been an administrator watcher for years, it would be very hard to get back the sabbatical leaves and the travel time. Teaching loads could easily go back to fifteen hours. You could be in a much worse situation two years down the road than you are now.

Watkins: If we want our teaching loads reduced, then from the administrative point of view, we are going to have to show them something for it. Right now we are paid to teach and to do nothing else, at least in my district, and I think that is common. The salary scale depends entirely on the number of years you have been there and the number of units you have completed in school. If part of our responsibilities are to involve things that are not quite so quantifiable as the number of hours in class, then part of the salary scale is going to have to be based on merit, and I just wonder how many community college instructors in this country are going to be willing to do that to the salary scale.

CURRICULUM REFORM

Curnutt: This discrete math furor worries me a bit. That there should be math majors, let alone people in user-type areas, who might never see calculus really scares the devil out of me. I don't like that idea at all. That's not to say that we couldn't do a lot towards upgrading the discrete side of things. My paper has a very modest list of things that we might incorporate into our courses that would upgrade that side of things. Many of those topics were in our curriculum at one time or another; standard college algebra used to have simple probability and counting types of problems. But that sort of disappeared with the elementary functions approach and I think that is a loss. There is probably room to fit those ideas into the existing courses. Peter Renz made the point in his paper about curriculum changes being evolutionary rather than revolutionary. Not only do I think that is the way things will go, but that's the way things should go.

Content may not be the real issue we have to deal with. For students who are not necessarily mathematics majors, there are issues more important than mathematical content--general problem-solving skills, heuristics in the Polya tradition, and ability to communicate clearly and logically. If we are going to pay more attention to these sorts of broader goals, we are going to have to pay a price, content-wise. I don't know how to pay this price, or what I am willing to give up for it, but there are much larger issues than just what topics ought to be in the mathematics curriculum.

It's been suggested that community colleges are a good place to experiment with curriculum change. Some types of community colleges might be. I'm not sure about transfer-oriented institutions. It's very difficult to experiment and design new curricula when you have restricted workloads. We've probably all been suffering the same thing over the last few years--severe budget restrictions. At Bellevue we have a heck of a time. If we get a class with less than 40 people, then we don't teach it. If you are going to experiment with new classes, you need flexibility. You are going to suffer a decrease in enrollment in those kinds of classes until they get off the ground. It is very difficult with those budget restrictions to spend any time experimenting with new things. Second, if you are a transfer institution, you are pretty much driven by what is required at the four-year schools; you've got to get people ready to go into those programs. Those two factors make it difficult to do any really serious, extensive experimentation at the community college level.

The biggest shortcoming I can see coming from teaching in a community college is the lack of variety of things I get to teach. Somebody who teaches at a university or a four-year college probably gets a chance to teach abstract algebra to juniors and seniors once in a while, or a two-quarter sequence in probability, or something like that. I'm pretty much constrained to do algebra through calculus, and every once in awhile I get a shot at differential equations or linear algebra. The most attractive thing to me about discrete mathematics is a selfish thing; I like something different to teach and something different to learn.

Page: It seems to me that two-year college faculty are being bypassed. I think that the money that is going into research in teaching at the elementary schools, secondary schools, and the colleges isn't being geared towards the enrichment and development of community college faculty. The research and grant money that goes to four-year people, who publish papers and articles, is fine because it enhances their own professional status. That kind of money is necessary to the actual professional survival of community college people. There is not enough time for community college mathematics faculty to do research, but that's not the only part. There is an even more difficult thing and that is continually teaching the same elementary subject matter to the weakest type of student. We talk about burnouts; I make the analogy in the sense of maybe being a gas station attendant who continually dispenses the same product. I never teach linear algebra. We don't have linear algebra; we don't teach differential equations; we don't teach probability. Those of us who are lucky get to teach a course in calculus. Now I am one of the world's greatest experts on the Pythagorean theorem, and I can do fractions very well. If it wasn't for a lot of other activities that I get into, there would be a deadening effect; there is a numbing there that most people are not aware of. They think it is just the amount of hours. The mind sets and one begins to think: "That is what I know." When I start to look at an article, it is intimidating. I don't remember this; there is a whole interplay. The national organizations have been remiss and are really not addressing the needs of the faculty of the community college. In fact, I would like to see AMATYC and the MAA take a stand and say that this component of the mathematical community is being bypassed. They have the greatest needs. As you point out, it would be a result of these developments that will lead to changes in the curriculum. But I think we are going about it in the wrong way. I think that those who teach have to be comfortable, feel inspired, and be enthusiastic, and everything else will follow from that--even from the self-image point of view.

Is Failure to Transfer an Indictment?

Sharp: If 50% of our students intend to transfer, and we transfer only 10% of them to four-year institutions, then that's a tremendous indictment.

Cohen: It's too easy to say that we are doing something wrong since more than half of the students intend to transfer, and only 10% ever do. You can think of the American system of education as a pipeline stretching from kindergarten to the doctorate. Three million people enter kindergarten, and we grant only thirty-five thousand doctoral degrees each year. There is leakage along the pipeline all along the way. A couple of generations ago most of the leakage occurred at the eighth grade; then it was the twelfth grade. Now, it is in the community colleges in grade 13 and 14. By the way, that has been a static figure for 20 years now. Is that an indictment? Must we castigate the community colleges for occupying a place in the system at grades 13 and 14 where in this generation most of that leakage occurs? I'm not ready to do that. It happens, it's true, that more than half intend to transfer, and only 10% do transfer. Must we say that something is wrong? That the community colleges are doing something bad? I can't handle that inference. What if we were to pass everybody who wanted to transfer and moved the leakage point up to grades 15 and 16, would that make us better in the community colleges? That's where you want to be careful about those inferences.

MATH--THE HEAVY, CURRICULUM DISASTERS AND PROBLEM SOLVING

Tucker: One really needs to go back to basics. If you just start with the basics, you can build something which is very different from what you have now and justify it about as well as you can justify what you have now. We need to look at common features, and the common feature that so many people have stressed is problem solving. We want students to understand what they're doing--to be able to talk about the mathematics, not to just have the techniques. One of the big concerns right now is the calculus-preparation orientation of college mathematics students. There are things in those courses about manipulations in algebra that are good for everyone. People don't, however, teach algebra thinking that they are preparing students to do computer programming. The texts aren't written that way. Everything is pointed towards calculus, and students ask why they need "this," and they are told they will need it if they take calculus in college. It turns out that most people take calculus in college as a screening device to test their mathematical maturity. So everything they learn up until then is really to prepare them to pass the mathematics test in college. The concepts of calculus that are used by an economics major or somebody in medical school are often very thin and intuitive. The whole thing is a screen. Math is often the heavy in this drama.

I know there are some engineering school administrators who have considered dropping calculus as the first college mathematics course and instead instituting some kind of problem-solving course to be taken before calculus. The usual assumption is that when you take calculus you already know how to solve problems. That is not always a valid assumption. There is a huge amount of material that people are supposed to master and then go on. It's all somehow a bit of a hoax because when you have a prerequisite for a course, in my experience, it's a course the students think they're supposed to take first. It's not a course they are supposed to learn something in. You ask the students: "Did you take such and such course?", and they say, "Yes, we took it." You then say, "Don't you recognize it?", and they say, "No." There is this barrier that has to be taken down. There is this game going on, and *learning* is something that gets washed out all along the way, and so the initial questions remain: what is a college, what is a university, and what is a high school? It is really all up for grabs.

Kaput: I fully agree that it's a large hoax. In fact, when you get beyond the point of giving these simple multiple-choice sorts of assessments--trying to measure the difference between the input of the behavior before taking and after taking the course, using lousy measures at both ends--you don't really discover that the students learn anything. You aren't going to learn anything yourself from those measurements. That whole enterprise, including the placement testing, predicated on a status quo, which is really a disaster, and the questions about faculty burnout, discrete versus applied mathematics and so on, are peripheral questions to the central issue of what students are really supposed to be learning. I would argue that they should be learning mathematics if we are going to be teaching mathematics, and that they should be learning to think mathematically. In some ways you can call that problem solving. If we are going to talk about mathematical problem solving, let me just call it _thinking mathematically_ because that is what mathematicians do, and that is what they are doing when they are applying mathematics--using it somehow to connect a wider world of experience. And so far we have not been doing that. The curriculum as it exists now doesn't teach that; it's a disaster. It teaches symbol manipulation, and it does that in an unmotivated way. The sequence that takes students through arithmetic, elementary algebra, intermediate algebra, college algebra and trigonometry, and calculus is just one long disaster. That fact is making itself evident in our discussion today, and in everything we see, and in everything we do in association with mathematics that has been taught at virtually all levels. We talk about faculty burnout; we talk about overloads. Consider how inefficient that system is that tries to teach students dead mathematics, to teach dead formalism over and over again to some of the same students in the same ways. Of course, it's going to be a disaster. No amount of money put into that particular system is going to save it. It's diseased; it's rotten at the core.

Rodi: I have an entirely different view of that. It seems to me that in all of human learning, be it mathematics, English, history, or anything else, there are certain central bits of information that people have got to be exposed to. Now, one way to do it is to have everybody go back and relive the learning experience from Plato to the present. Most of us don't have time in our lives to do that. Another way to do it is to synthesize in some fashion some of the wisdom of the ages and try to give it to people in an efficient way. Many of these courses in math, English, history, philosophy, or anything else are aimed at giving people in the year 2000 a certain amount of the accumulated wisdom of the past 2000 years. Now, that doesn't mean their education has got to stop there. Nor does it mean that they shouldn't develop all sorts of other skills, but I am worried about throwing out the baby with the bath water. Humankind has learned something in 2000 years, and we ought to find some effective and efficient way of transmitting that to the bulk of our populace. We may not have the most efficient way, but I am not ready to go so far as to say that we are not transmitting, by and large, valuable and useful information and skills to the vast bulk of our populace in the method that has been developed.

Renz: Let me come to the defense of the mathematics community, defending it against the onslaughts that Jim has made. I think one of the reasons that the people around this table might regard mathematics education as a failure is that many of the students don't learn or don't retain specific bits of information that we think are important to us as mathematicians. But those people who do go on to business and other fields do retain to some extent an ability to deal with matters quantitatively. In more than ten years in publishing, I think I made use of a geometric series once, and I used the derivative on two occasions, and that was about it. You don't need fancy mathematics for most business purposes. I am more quantitatively inclined than most people, and I still found myself hard-pressed to use any kind of fancy mathematics at all. All of those optimization problems that you work out don't really make any sense because no one really knows what the supply and demand functions are. Prices are negotiable, markets are very difficult to get your fingers on. Optimization techniques give you conceptual tools which people can use even though they can't actually work out the answers. There are areas where you can use optimization. If you are running an oil refinery, you can use sophisticated mathematics. But in ordinary business, this is not possible. I think people also learn about problem solving, which is really a very large and general kind of ability. But much of problem solving is not the sort of thing that you teach people in mathematics courses where you have very circumscribed problems. You have a very definite objective. One of the principal means for solving problems in real life is to change the problem. In the case of trying to figure out how one can make money publishing a book, sometimes the solution is not to publish it. Maybe the solution is to regard it as three books instead of one. These are all possibilities, and changing the context of the problem is always open to you. So, in summary, I don't think that you should fall into despair simply because students don't have a mastery of specific techniques.

PROGRESS AND MONEY, MONEY, MONEY

Ellis: If we are going to have a big thrust in computers, all instructors should have computers, or computer terminals on their desks right now. We must have them, and I'm not going to pay for them. I refuse to do that in the face of this tremendous demand for computing/mathematics skills from the outside world. It shouldn't come out of my hide, but it already came out of my hide because I don't make as much as those folks outside academe.

Gordon: One of the biggest problems faced by two-year schools throughout the country is a lack of resources, usually money. Our library facilities range from poor to dismal. I am at a school with 23,000 students, and they just cut back to two math journals. Grant offices are almost non-existent; on many of the campuses there are no people who can assist people. There is a small pool of talent or experience on the campuses in terms of giving suggestions about how to improve grant proposals, how to go after them, where the grant money is in terms of information on computers, etc. Many of the two-year faculty just don't have access to the information on how to obtain money to attend conferences, small in-house grant projects for instructional development, and curricular development. None of those things are available to the two-year schools. Somebody mentioned seminars;

there is no money to bring people in. More than anything else there are just no resources available for information on all of these things. This is an area where national organizations should help.

Page: What I seem to be gathering is that there are no proposals going forward because two-year people can't write the proposals. Well, the high-school teachers aren't that much stronger at writing proposals either, and so the point is, we don't have to predicate everything on the fact that the two-year people have to become experts at writing proposals. Why can't these proposals come from some other sources as well, such as four-year people? The point I'm making is that funding should be directed towards the two-year people and not be withheld until they develop the opportunity to go after it.

Albers: In the case of high-school proposals, they are often done in conjunction with a university's school of education. These are team efforts. We have not seen two-year college/four-year college efforts or two-year college/high-school efforts.

Ross: The first step in writing proposals is to keep the F.O.W.G.I. principle in mind (F.O.W.G.I. = Find Out Who's Got It). Basically, NSF has functioned on the trickle-down theory and has not funded

K through 12. They fund universities, and that's where the money comes from.

Fusaro: The two-year college is like the invisible man as is the four-year college, as far as NSF is concerned. You can pass out all the information you want, but the money is not really there right now for two-year colleges. NSF money right now is for university research and high-school projects. If the two-year colleges can tie into that, you can get it. You can submit all the grant applications you want, but the cash is not there. In fact, the Society for Industrial and Applied Mathematics--I am Chairman of their Education Committee--which really has not been too active in education matters, is suddenly going to high schools and has become very interested in high schools. I wonder why?

Kaput: NCTM has a proposal that I suspect will be funded shortly to create a cadre of microcomputer experts. They have assembled the best people that they were able to find who can teach people over a period of about five or six days, using the best materials and a variety of computers, about the use of micro-computers and the teaching of mathematics. Now, the assumption is that the people who are going to come to the NCTM are coming from various regions around the country, who themselves will become work shop leaders for teams that will then work with particular high-school districts. You could define a particular change objective, and find a group of experts, who could then train other trainers in a systematic way, and who could then work with individual community college systems. This highly leveraged expertise can pay off. You don't use a lot of money because you are not paying a lot of people at the center of the operation. You're also gaining access to a lot more money because there may be local and state money available if you come up with a decent proposal. There are various sources around that can be had if you have a very particular idea, and if you are basking in the glow of some national organization that is associated with the effort and has blessed it in some sort of way.

Rodi: Are community colleges the right place to look for leadership in curriculum change? Larry Curnutt looked at it from the point of view that we pretty much follow the coattails of four-year schools. I entirely agree with that. I don't see us saying that we are going to have curriculum change and the universities saying, "Oh, yes, we will now change our curriculum to reflect what you are doing." I can see it working the other way around--the universities saying, "We're having curriculum changes, and now you are going to have to change." Ideally, I see us getting together and talking about what would be cooperative curriculum changes. There are other things which are necessary for curriculum change besides expertise and money, and we don't have those either.

Albers: It has been suggested that a new round of NSF institute programs is needed to help facilitate curriculum development. Some people at NSF say the high-school institutes were a failure, and for the most curious reasons; they didn't upgrade the quality of high schools. Many of those high-school teachers who attended NSF institutes for a few summers or for an entire year became our two-year college mathematics teachers. They left their high schools after upgrading their knowledge. Two-thirds of our faculty previously taught in secondary schools. The situation today is very different. If you say you want an institute program for two-year college faculty, I don't think you're going to be talking about moving those people up to another level.

Part 2.

TECHNICAL MATHEMATICS

TECHNICAL MATHEMATICS: A DILEMMA

by

Keith Shuert

ABSTRACT

TECHNICAL MATHEMATICS: A DILEMMA

Keith Shuert

This paper addresses the problem of how community colleges can effect an interface between the technical faculty and the math faculty with regard to appropriate mathematics education for the vocational/technical students.

A process for the development of a tech math sequence of courses is offered which involves the input of both math faculty and technical faculty on a cooperative basis. The initial thrust should come from the career faculty. After both groups mutually arrive at a list of math topics, the math faculty should formally develop the course and submit it to all the involved faculty for final approval. A number of possible resolutions are given should this cooperative venture break down.

The issue of who should teach technical mathematics is discussed. Instructors should have experience in industry and have completed math courses through calculus. The view is expressed that it is probably easier for technical faculty to upgrade their math than for the math faculty to become familiar with the appropriate applications of tech math.

A list of objectives for the first of a two course sequence in tech math is provided. The course content was arrived at by consultation between the math and the applied technology faculty at a community college.

A review of related literature is included under the headings of cooperation between mathematics and occupational faculty, technical math competencies, and technical math curriculum development.

THE PROBLEM

In most community colleges, technical mathematics is offered through departments of mathematics and is taught by faculty who have degrees in mathematics and who have a liberal arts bias. They usually have little knowledge of the kinds of applications their students will meet in the various technical courses they will be taking. The math faculty tend to rely on any examples their text provides but have little actual first hand experience in the technical fields from which their students come. These math faculty certainly can prepare their students with the algebraic skills they might have to use in their technical fields; but there is no guarantee that they are giving the students the specific skills and training that the technical education faculty want their students to have.

On the other side the technical education faculty find their students poorly prepared to handle the mathematical requirements of their courses. They approach the math faculty saying that they need help. They delineate the math skills their students need for their particular discipline. The trouble is they never seem to get what they request. This situation puts the technical faculty in a bind. They can continue to beseech the math teachers to provide their students with the prerequisite math skills, with little evidence of return. They can try to take valuable classroom time out of the technical classes to teach these skills to their students. Or if their college offers apprenticeship programs, they can turn to apprenticeship math courses, which may be called an Introduction to Algebra or even Algebra, but which often are just glorified shop math courses which do not cover algebra with sufficient rigor to be applicable to a wide range of different vocational technical programs.

If technical faculty approach math faculty with their list of wants, math faculty tend to look down their list of offerings to find the course that happens to come closest to the want list. This is usually a Tech Math I and II sequence. The problem is that most math faculty teach Tech

Math I and II as though they are Beginning and Intermediate Algebra courses, mainly by default, due to their lack of experience with the technical fields they are supposed to be serving.

The problem then is how can the technical faculty and the math faculty cooperate to design an appropriate mathematics education for the vocational/technical students.

A PROPOSED SOLUTION

In most cases a new Technical Mathematics sequence should be developed. In some community colleges the traditional tech math courses are no longer offered or are experiencing declining enrollments in spite of increased vocational/technical enrollments requiring math. The enrollments have shifted to apprenticeship math classes, or the standard Beginning and Intermediate Algebra courses.

The successful development of a new tech math sequence should proceed as follows:

1. It should be recognized that the development of a new tech math sequence, usually two courses, is necessary to break away from the built in, long standing biases of both math faculty and career education faculty.
2. This development should start with a series of meetings between the vocational education faculty and the math faculty. Those attending these meetings will be determined by the dynamics of any given institution. It may be department heads, discipline representatives, departmental committees of those interested, or all the vocational/technical faculty in programs requiring math with all the members of the math department. Certainly, the more involvement by faculty the greater chance of a successful outcome.
3. Together these two groups of faculty should come up with a list of topics to be covered in this sequence. There should be a free flow of ideas and an open and exhaustive exhange of possible topics. The initial

input and thrust should come from the vocational/technical faculty. The math faculty should accept these suggestions and expand on them, questioning precisely what math skills are needed and suggesting related skills in an attempt to pin point what is really needed, and to insure that all related and requisite skills are included. The career faculty should be as detailed in their list as possible. The math faculty should be sure all requisite skills are included but only those that are necessary to meet the requested skills.

4. The bulk of this interaction and exchange can occur at one meeting, but there should be follow up meetings to refine and polish the final list of topics.
5. The math department should then take the responsibility of formally developing the tech math sequence in accordance with the institution's curricular procedures. They have the expertise and experience to insure the sound development of these skills.
6. These courses should be submitted to all the faculty involved for final approval. This should include the vocational/technical faculty in those disciplines or programs requiring math and the math faculty. It is often at this step that the process come to a screeching halt. This sometimes occurs because all the affected faculty, especially the vocational/technical faculty, were not involved in the original deliberations. This is severely compounded if the institution is a multi-campus institution with vocational/technical faculty and math faculty at each of several campuses. Breakdown will also occur if there has not been enough dialogue in the original meetings or if the exhange has not been sufficiently open and free of bias. The key to sucess is tolerance by each group of faculty for the views and position of the other group. The math faculty are concerned with the students' preparation for subsequent courses. The vocational/technical faculty are primarily interested in students being able to handle work applications.

However, the real threat to success and the one factor that will sabotage these efforts faster than anything is strong reluctance by either group of faculty to give up or lose student enrollments in their courses. If the math department has had enrollments in their traditional tech math sequence, they may find it difficult to give up their liberal arts bias and admit that they have not been offering tech math in the most effective way. Chances are, however, that most math departments have all the students they can handle. Their problem is most likely having too few full-time faculty and experiencing great difficulty in hiring enough part-time faculty, especially well-qualified ones, to cover all the sections they are offering. So, if the technical faculty balk at this stage, the math department most probably will discontinue its involvement.

The vocational/technical faculty are the most likely to halt the proceedings at this approval stage. Often they have been disappointed by the level of math skills of their students and have blamed the math courses for this lack of preparation. They have often chosen to fall back on their own apprenticeship math courses in lieu of the tech math courses and have enjoyed the accompanying increase in enrollments. At this juncture they will resist any attempts to supplant the apprenticeship math courses by the new tech math courses. This log jam must be dealt with head on if it occurs. Some possible solutions are:

(1) Discuss this issue from the onset and come to a mutually agreeable resolution before the development of a new technical math sequence is even started. It could be resolved by deciding which courses will require apprentice math and which will require the new, to be developed, technical math. This could be a natural result of preparing the list of required math skills for the various career courses. Those courses which require math skills that are covered in an apprentice math course would use that course. Those which better fit the technical math sequence would require the technical math course. There is obviously no place for selfish

bias in these determinations, but an honest and genuine concern for the needs of the students and for the best way to meet those needs.

(2) Resolve the conflict by establishing a mutually agreeable procedure to determine who teaches the new technical math sequence. For example, the Tech Math courses could be team taught. A math instructor would cover the skills and a vocational/technical instructor would cover the applications of the skills. Or, the technical math course could be crosslisted in both the math department and the vocational/technical department. The enrollment counts could then be split between both departments.

(3) Solidify the distinction between apprenticeship math and technical math by restricting apprenticeship math to apprenticeship programs, and require two-year vocational technical programs to use the technical math sequence. This solution will not likely please all the faculty, especially the apprenticeship faculty, particularly if the apprenticeship enrollments are on the wane.

(4) Decide to place the total responsibility for technical math with the vocational/technical/apprenticeship training department(s). After all, most math faculty are not well versed on the practical applications and uses of the mathematical concepts and skills that they teach. They usually prefer to stress the purely mathematical aspects of their discipline. This may be a workable solution in many instances but is not the most desirable one. There is too much that is lost by the students, the faculty and the institution if this opportunity to bring about cooperation between vocational technical faculty and math faculty is avoided. The vocational/technical faculty may not be qualified, or interested in teaching technical math. The apprenticeship courses may not meet the needs of all the career courses, shortchanging a number of students. The math faculty would miss the opportunity to expand their expertise and effectiveness as math teachers. Most of them are aware of the deficiencies in their background

relative to technical math applications, and trust that strong skill development in algebra will see their students through their vocational/technical courses. Any institution certainly benefits from a healthy mutually supportive interface between vocational/ technical faculty and liberal arts faculty. Administrators search for ways to build these bonds and counteract the differences in ideas, values, opinions, and respect that seem to keep vocational/technical faculty and liberal arts faculty apart.

WHO SHOULD TEACH TECHNICAL MATHEMATICS?

An ideal way to staff technical math courses is to have the vocational/technical faculty and the math faculty get together and agree on qualifications of any faculty member wishing to teach technical math. They could form a joint committee to review the qualifications of anyone wishing to teach technical math. The department in which the technical math courses are housed would then select from those who are declared qualified by the joint committee. Ideally, instructors should have experience in research and development in industry and have completed math courses through calculus. Certainly, instructors should be disposed to emphasizing an applied rather than a theoretical approach.

This will require preparation by both groups of faculty. Technical faculty would most likely have to update their mathematical skills before they would be prepared to teach technical math. The math faculty, on the other hand, would have to acquaint themselves with the various applications of math in the programs which the technical math courses serve. It is probably easier for the technical faculty to upgrade their math than for the math faculty to learn the appropriate applications.

A TECHNICAL MATH COURSE

Below are the objectives for the first course of a two course sequence in technical mathematics which might be called "Math For Technicians I." It is intended to be a four credit hour course. The course content was arrived at by consultation between Math faculty at one campus of a four campus community college and the Applied Technology faculty at the same campus. The recommended text for this course is Introduction to Technical Mathematics, second edition, by Allyn J. Washington, published by the Benjamin/Cummings Publishing Company, Menlo Park, California.

Catalog Description

Review of arithmetic, including fractions, decimals, percents, and ratio and proportion. Calculation with approximate numbers, and rounding off. Metrics, measurement conversion. Operations with signed numbers and algebraic expressions. Factoring and algebraic fractions. Solving linear equations and inequalities, and literal equations. Exponents, roots, and radicals, including scientific notation. Perimeter, area, and volume of basic geometric figures. The use of calculators in the solution of technical and industrial problems will be emphasized.

Design Criteria and Performance Goals

DC I: The student will solve, without the use of a calculator, arithmetic problems in whole number, fractions, decimals, percent, and ratio and proportion.

- PG I: The student will solve problems in adding, subtracting, multiplying, and dividing whole numbers.
- PG II: The student will solve problems in adding, subtracting, multiplying, and dividing fractions and decimals.
- PG III: The student will solve problems in percent, including finding percentage, base, or rate, given any two of the three.

PG IV: The student will solve problems in ratio and proportion, including finding the ratio of two numbers, dividing a number into a given ratio, and finding the missing number in a proportion.

DC II: The student will carry out calculations involving approximate numbers and rounding off of approximate numbers.

PG I: The student will compute with approximate numbers, using the arithmetic operations.

PG II: The student will determine the appropriate number of decimal places in an approximate calculation.

PG III: The student will round off approximate numbers to the appropriate number of decimal places.

DC III: The student will become familiar with the metric system, and will carry out conversions in both the metric and English systems, and conversions between the systems.

PG I: The student will become familar with mass and length units in the metric system.

PG II: The student will convert between units in the metric and English systems.

PG III: The student will convert units between the metric and English systems.

DC IV: The student will carry out, without the use of a calculator, the arithmetic operations with signed numbers and algebraic experessions.

PG I: The student will add, subtract, multiply, and divide signed numbers.

PG II: The student will add, subtract, multiply, and divide algebraic polynomials.

DC V: The student will factor algebraic polynomials.

PG I: The student will factor polynomials using the common factor technique.

PG II: The student will factor polynomials as the difference of two squares.

PG III: The student will factor trinomials as the product of two binomials.

PG IV: The student will factor polynomials using combinations of the above techniques.

DC VI: The student will solve linear equations and inequalities in one variable, and will solve literal equations for an indicated variable.

PG I: The student will solve linear equations that do not involve fractions.

PG II: The student will solve linear equations involving fractions.

PG III: The student will solve and graph on the number line linear inequalities involving one or more operations.

PG IV: The student will solve literal equations for any variable in terms of the other variables.

DC VII: The student will carry out simplifications and arithmetic operations with exponents, radicals, and scientific notation.

PG I: The student will carry out the arithmetic operations with algebraic expressions containing positive, negative, zero, and fractional exponents.

PG II: The student will simplify radicals containing arithmetic and algebraic expressions.

PG III: The student will carry out the arithmetic operations with radicals.

PG IV: The student will convert between standard and scientific notation.

PG V: The student will multiply and divide numbers in scientific notation.

DC VIII: The student will carry out computations involving basic geometric figures.

PG I: The student will find perimeters and areas of plane geometric figures.

PG 11: The student will find the surface and volume of solid geometric figures.

REVIEW OF RELATED LITERATURE

Cooperation Between Mathematics and Occupational Faculty

Very little has been written on the subject of articulation between mathematics faculty and vocational technical faculty. Davis conducted a study to determine the level of cooperation between mathematics and occupational-technical faculty in the development and delivery of occupational-technical mathematics courses at 100 randomly selected two-year colleges. Five key factors were identified as influencing the level of cooperation: (1) the quality of informal communication; (2) the process for the development and review of the content for occupational mathematics courses; (3) the departmental location of responsibility for occupational math course instruction; (4) enrollment; and (5) the age of the college. The study noted with concern four trends observed as impacting upon mathematics education. (1) Mathematics service courses are increasingly being taught by faculty outside mathematics programs. When this occurs, mathematics faculty are very likely to have no involvement in the development and delivery of those courses. (2) A substantial minority of mathematics faculty still retain the view that "career" programs are inferior to "academic" programs. (3) Limited numbers of mathematics faculty have interest in teaching mathematics service courses. (4) Limited numbers of mathematics faculty have confidence in their ability to teach mathematics service courses.

Technical Math Competencies

Anyone developing or revising technical math courses needs to determine the math skills and competencies which are required for the technical programs they service. Some work has been done to accomplish this for various occupations.

The Northeast Wisconsin Technical Institute in Green

Bay, Wisconsin surveyed employer and employee expectations of and satisfaction with the mathematics competencies essential to perform in fifteen occupations. Over 1800 graduates of fifteen occupational programs in two Vocational Technical Adult Education districts in Wisconsin and their employers were surveyed by mail. A set of mathematics competencies for these occupations resulted. This led to the development of a profile matrix of mathematics competencies which lists those competencies common to all occupations and those peculiar to the fifteen specific occupations. Their report contains a master list of 600 mathematics competencies.

Samuel Self conducted a research project to develop an applied or technical mathematics curriculum which would meet the needs of vocational/technical students at the community college level. Staff members from 10 community college in Texas participated in the survey to help determine and validate job competencies for the occupational areas of diesel mechanics, auto mechanics, radio/TV repair, air conditioning, welding, machine shop, printing, drafting, and electronics. The project results include a list of the mathematical concepts requisite for entry-level competencies in each of the selected occupational areas, a set of structured, sequential technical mathematics units designed to meet the needs of vocational/technical students in the selected occupational areas, curriculum guides for each of the technical mathematics units, self-instructional learning packets for each of the technical mathematics units, and performance-based pre-and post-tests for each of the technical mathematics units.

Edwards and Roberson surveyed 470 graduates of the six engineering technology programs at Wake Technical Institute - Architectural, Chemical, Civil Engineering, Computer, Electronic Engineering, and Industrial Engineering Technologies - and 227 of their employers to determine the science and mathematics topics most needed by engineering technicians. The survey contained 81 items under the following 17 topical areas: mechanics, fundamentals of electricity/electronics, light, sound, heat, modern physics, chemistry, biology, geology, data processing, algebra, trignometry, logarithms, geometry, analytical geometry, calculus, and statistics. The survey revealed that mathematics topics were important to all graduates, various science topics were needed for different technology areas, knowledge of computer language was important to all but architectural technicians, and science and mathematics topics were more important to graduates than to employers.

Barlow and Schill did a study to determine the kind of mathematics the electronics technical worker actually used or needed on the job. The data were gathered from 90 technical workers from 44 California electronics companies, and from 29 instructors from 45 junior colleges with electronics programs. The mathematical concepts or skills defined as essential to electronics technical workers in research and development were conversion of fractions to decimals, percent calculations of tolerance, changing percentage to decimals, conversion of metric to the American measuring system, square roots, division of signed numbers, scientific notation, estimation of arithmetic problems, multiplication and division with exponents, raising to a power with exponents, use of negative exponents, ratios, and the Pythagorean Theorem.

Newton conducted a project to identify the basic and functional mathematics skills that metal trades students needed to meet entry-level job requirements as defined by the metal trades industry for the following occupations: combination welder apprentice, machinist helper, precision metal finisher, and sheet metal worker apprentice. He provides pretests on these skills as well as student learning projects that prepare metal trades students to read, understand, and apply the required mathematics and measuring skills.

Technical Math Curriculum Development

A number of people have been involved in the development of instructional material for technical math courses.

Queensborough Community College has a handbook designed to assist teachers of technical mathematics in developing practical curricula for their students. The underlying assumption is that technology students' needs are oriented to the concrete. The handbook describes the nature, scope, and content of curricula in Electrical Technology, Mechanical Technology, Design Drafting Technology, and Technical Physics, with particular reference to the mathematical skills which are important for the students, both in college and on the job.

Doversberger did an evaluation of the two-year engineering related technology curricula at Illinois Central College to determine appropriate course content for technical mathematics and physics, the level of theory to be taught, instructor qualifications and the appropriate department for offering these technical courses. It was found that specialized courses in mathematics and physics for technology programs should emphasize an applied rather than theoretical approach, that instructors teaching these courses should be sympathetic to technical education and have had industrial experience, and that technical mathematics and physics courses should be offered in the technology department.

Greenfield developed twenty Algebra-Packets for technical students at the community college level. Each packet contains a statement of rationale, learning objectives, performance activities, and performance tests.

Ellis and Payez developed an alternative method for teaching mathematics to occupational/technical students at Paul D. Camp Community College in Virginia. Recognizing the need for the student to see the relevance of his or her studies to his occupational goals, an effort was made to structure mathematics courses to fit specific student needs within each occupational program. Mathematical skills were broken down into groups, and the skills essential to each program were determined. Common blocks of mathematics instruction in fractions, decimals, etc., were supplemented by problems tailored to and written in the individual occupational jargon.

Melton developed a study guide to train technicians in the use of electronic instruments. The guide provides that part of the mathematics content related to algebraic and trignometric equations and their applications. The following topics are included: (1) linear equations in two unknowns; (2) trignometric equations and vectors; (3) systems of linear equations; (4) quadratic equations; (5) complex numbers - imaginary roots of quadratic equations; (6) equations containing fractions; and (7) exponential and logarithmic equations.

The Technical Education Research Center at Waco, Texas has prepared eight modules designed to strengthen mathematical and laboratory skills in the areas of units, graphing, logarithms, dimensional analysis, and basic trigonometry. The modules are titled: "Formula Interpretation"; "International System of Units"; "Logarithms and Exponents"; "Angles and Triangles"; "Vectors and Scalers"; "Dimensional Analysis"; "Reading and Drawing Graphs"; and "Precision, Accuracy and Measurement."

Flannery wrote a textbook to accompany a series of video tapes which provides information, examples, problems, and solutions relating to mathematics and its applications in technical fields. Of special interest is Chapter VI which considers the applications of mathematics to plane surveying, computer science, machine parts inspection, automotive technology, aviation, manufacture engineering processes, pharmacology, and optical technology.

The Oregon Vo-Tech Math Project provides verbal problem sets for occupational courses offered in Oregon Community Colleges. The problems were selected to provide the math skills necessary for job entry and performance in several vocational/technical occupations. An individualized format permits students to work problems in various career areas.

SELECTED BIBLIOGRAPHY

Barlow, Melvin L. and Schill, Williams J. "The Role of Mathematics in Electrical-Electronic Technology." Research Report, California University, Los Angeles, 1962.

Davis, Ronald Marshall. "The Development and Delivery of Mathematics Service Courses in Two-Year Colleges." Doctoral Dissertation, University of Maryland, 1980.

Doversberger, Betty, "An Analysis of the Practices in the Teaching of Technical Mathematics and Technical Physics in Illinois Junior Colleges." Seminar Paper, August, 1970.

Edwards, Timothy I. and Roberson, Clarence E., "A Study to Determine the Basic Science and Mathematics Topics Most Needed by Engineering Technology Graduates of Wake Technical Institute in Performing Job Duties." Research Report, Wake Technical Institute, Raleigh, N.C., 1979.

Ellis, Harry and Payez, Joseph F. "A non-Traditional Method for Teaching Mathematics to Occupational/Technical Students. Paper presented at the Virginia Sectional Meeting of the MAA, Rockville, Maryland, November, 1974.

Flannery, Carol A. Technical Mathematics: Restructure of Technical Mathematics. Mountain View College, Dallas, Texas, 1982.

Greenfield, Donald R. "Condensing Algebra for Technical Mathematics." Doctoral dissertation, Nova University, 1976.

Hannon, Ralph H. "In Plant Programs Involving Technical Mathematics." MATYC Journal, 10, 3, (1976), 159-160.

Loiseau, Roger A. "A Study of Coordination Between Mathematics and Chemistry in the Pre-Technical Program." Doctoral Dissertation. Nova University, November, 1974.

Melton, Roger H. Algebraic and Trignometric Equations With Applications. Saint Louis Community College at Florissant Valley, Missouri, 1976.

Newton, Lawrence R. "Shop Math for the Metal Trades." Project Report, Weber State College, Ogden, Utah, July, 1981.

Self, Samuel L. "Community College Technical Mathematics Project." Research Report, Texas A and M University, December, 1975.

Wiggins, E. Foster, "Survey of Technical Mathematics Instruction in the Community Colleges of Massachusetts." MATYC Journal, 11, 2, (1977), 96-99.

"Identification of Mathematics Competencies for Vocational, Technical and Adult Education through a Survey of Employer/Incumbant Employee Expectations." Research Report, Northeast Wisconsin Technical Institute, Green Bay, Wisconsin, August, 1980.

(DISCUSSION BEGINS ON P. 131.)

OCCUPATIONAL EDUCATION AND MATHEMATICS: OWNERSHIP MAKES THE DIFFERENCE

by

William Warren

and

James R. Mahoney

SUMMARY

To develop and maintain the appropriate relationship between community college mathematics courses and technical education curricula requires an understanding of institutional mission and the fit of specific programs into that mission. The community college philosophy which was best conceptualized by President Truman's Commission on Higher Education in 1947 revealed essential characteristics to make community colleges truly a "community" resource, and called for the integration of general education and vocational/technical education.

Within this context, the development of appropriate and acceptable skills in the use of mathematics requires a commitment to both course content and standards of achievement by the institution, technical faculty, general education faculty, and employer groups. Without this commitment -- or feeling of program "ownership" -- groups with varied vested interest find themselves at odds with each other in setting standards and course requirements.

By working together to establish standards, evaluating incoming students, specify remediation, and maintain standards to determine academic success, a partnership will evolve to serve the institution and students in a more responsible fashion.

OCCUPATIONAL EDUCATION AND MATHEMATICS OWNERSHIP MAKES THE DIFFERENCE

William C. Warren
James R. Mahoney

A discussion of the relationship between community college general education math courses and the technical education curricula that they support, with a stress on problems which are evident in that relationship and some possible solutions, requires first an understanding of institutional mission and the fit of specific program efforts within the mission. Therefore, a sketch of the history and emphasis of community colleges by which we can gain that perspective is necessary.

The term "community college" was coined officially by President Truman's Commission on Higher Education in its report issued on December 11, 1947. From that time on the terms "junior college" and "vocational/technical institute" described inadequately institutions that were designed to serve the communities in which they are located. The vision of the Commission focused on the essential characteristics of community colleges -- that they function truly as a community resource and that they "belong" to the community. Also, the Commission recommended that in addition to offering the first two years of a four-year or professional degree, the community college should:

1. Conduct community surveys to determine community needs.
2. Offer programs designed to meet the needs of a cross section of the population, including older adults who need to alternate their time between work and college attendance.
3. Integrate general education and vocational/technical education.

4. Serve as a center for a comprehensive program of adult education. (Vaughan, pp. 22-23)

These essential characteristics are as relevant and appropriate now as they were when they were conceived in 1947. They underscore what is central to community college activities toda We, as a community college system, continue to survey our communities, offer programs to meet the needs of the diverse individuals who attend our institutions, attempt to integrate general education into our vocational-technical education programs, and offer comprehensive programs for adults. The four characteristics identified by the Commission are primary to the community college mission as is the more traditional transfer characteristic. Discussions on the relationship between mathematics and technical education in the community college requires that these characteristics be understood as the basis upon which solutions must be devised.

The National Commission on Excellence in Education cited 13 indicators of risk in their report A Nation at Risk. The following eleven are relevant to the area of mathematics as they are to other areas of general education:

> International comparisons of student achievement, completed a decade ago, reveal that on 19 academic tests American students were never first or second and, in comparison with other industrialized nations, were last seven times. . . .
>
> Average achievement of high school students on most standardized tests is now lower than 26 years ago when Sputnik was launched.
>
> Over half the population of gifted students do not match their tested ability with comparable achieve ment in school.
>
> The College Board's Scholastic Aptitude Tests (SAT) demonstrate a virtually unbroken decline from 1963 to 1980. Average verbal scores fell over 50 points and average scores dropped nearly 40 points.
>
> College Board achievement tests also reveal consisten declines in recent years in such subjects as physics and English.
>
> Both the number and proportion of students demonstrating superior achievement on the SAT's (i.e., those with scores of 650 or higher) have also dramatically declined.
>
> Many 17-year-olds do not possess the "high order" intellectual skills we should expect of them. Nearly 40 percent cannot draw inferences from written materials; only one-fifth can write a

pursuasive essay; and only one-third can solve a mathematics problem requiring several steps.

There was a steady decline in science achievement scores of U.S. 17-year-olds as measured by national assessments of science in 1969, 1973, and 1977.

Between 1975 and 1980, remedial mathematics courses in public 4-year colleges increased by 72 percent and now constitute one-quarter of all mathematics courses taught in those institutions.

Average tested achievement of students graduating from college is also lower.

Business and military leaders complain that they are required to spend millions of dollars on costly remedial education and training programs in such basic skills as reading, writing, spelling, and computation. The Department of the Navy, for example, reported to the Commission that one-quarter of its recent recruits cannot read at the ninth grade level, the minimum needed simply to understand written safety instructions. Without remedial work they cannot even begin, much less complete, the sophisticated training essential in much of the modern military. (pp. 8-9)

Although this disturbing report addresses public secondary education primarily, the reactions it will cause will greatly impact programs in higher education. National discussions, deliberations, and decisions on new directions to be taken in response to the recommendations must be holistic, folding in the interests of mathematics teachers, students, and employers in both the public and private sectors. They must also be sensitive to the institution's financial resources. Our directions must be responsive to community needs as a whole with the concerns of the students and graduates foremost in our decisions.

The process of interaction between mathematics departments and departments of vocational-technical education involves at least four areas:

1. Establishment of standards (both entrance and exit competencies).
2. Evaluation of incoming students (relative to the entrance competencies).
3. Specification of remediation (for those not qualifying for unconditional entrance).
4. Maintenance of standards in determining academic progress.

STANDARDS

Standard-setting with a national perspective is a serious problem because of the sizeable numbers of colleges and their diversity. There were 1,219 community, technical, and junior colleges listed in the 1983 Community, Technical and Junior College Directory with state-wide student enrollments ranging from 1,201 to 1,221,430 and institutional enrollments ranging from a low of 132 full-time equivalent students to a high of 30,820. Setting uniform national standards that satisfactorily account for such diversity would take genius and ingenuity that is rare in education today.

Other issues that complicate standard-setting include:

1. Seconday schools' concerns that moving from open admissions to the setting of admissions standards will keep out applicants who would gain from a two-year college experience.

2. Instructional departments' (including general and occupational education) fears that the lack of standards will lead to erosion of a well-earned reputation.

3. Employer concerns that graduates may be certified with a diploma, certificate, or degree, yet, not be competent in the selected field of employment.

Pertinent to these concerns and fears is the institutional, departmental, and faculty perception of "ownership" -- that is, the responsibility an instructor, department, or institution (as well as any of the other constituencies -- such as advisory committees) accepts for the success or the deficiencies demonstrated by the graduates of a program.

Vocational/technical education historically has had a special burden to carry and has had to justify its existence more than any other curricula area. Institutions that offer postsecondary vocational/technical education funded in part by federal vocational funds are continually active in writing grant proposals, establishing technical advisory committees, filing state and federal end-of-the-year reports, and carrying out follow-up studies of their graduates. Vocational/technical department faculty become intimately involved in this process. Also, they work closely with students and employers. When a graduate from a vocational/technical department fails to exhibit the skills, knowledge, and attitudes expected by his/her new employer, it is the vocational/technical department and faculty as well as the graduate that bears the blame, that suffers the indignity. The institution is viewed negatively, the faculty is viewed deficient, or worse, and the department is, perhaps, viewed as being out of step with employers. Generally, the problems are brought to the attention of the institution and corrective actions are taken quickly. However, the most immediate and positive corrective action will be taken when "ownership" is accepted by all those involved.

Another term related to "ownership" to explain what I have in mind here is "partnership". One of the small ironies that I see in the rhetoric of community colleges today is that we tend to apply our best selves, our greatest energies, our best thoughts to relationships with individuals, programs, and organizations in the community. We do not apply ourselves nearly as rigourously to our own in-house operations. In a way, our institutional behavior is like that of a person who is marvelously personable with strangers and acquaintances but horrid with his/her own family. That is, we speak a great deal about building partnerships with local business, with labor unions, with professional associations, with churches, and with other power groups in the community -- but we do not apply the same drives to our in-college operations. I feel very strongly that we need to be just as helpful and service oriented -- just as cooperative -- within the college as we are within the community. My use of the terms "ownership" and "partnership" implies these behaviors.

To establish particular program standards requires that programs be developed with performance standards acceptable to both employers and the institution. Greer (1967) in a study of technical education described a conceptual model that recognizes the overlapping functions of institutions with an occupational education mission (the term "occupational education" includes both vocational and technical education). This broad occupational spectrum -- from a single skill occupation to multi-skilled engineering technician -- results in state, regional, and local program differences of significant proportion and interpretations of standards that cloud the issues in bringing about relevance to general education courses. These differences in programs and standards further complicate the efforts to bring about positive changes.

As can be seen in Figure 1 following, the truly comprehensive community college covers the entire occupational spectrum below the engineering technician or para-professional, whereas the vocational-technical school (vocational-technical institute as considered here) covers a narrower spectrum of postsecondary programs, and the technical institute has an even more narrowly defined function. The fourth category in Figure 1, the vocational-industrial or area vocational school, is at the secondary level and is omitted from this discussion.

Figure 1

OVERLAPPING FUNCTIONS OF DIFFERENT TYPES OF SCHOOLS

Training Objectives			
Single Skill Occupations	Skilled Craftsman	Technical Specialties	Engineering Technicians

```
                                                 Technical Institute
                                                XXXXXXXXXXXXXXXXXXXX
          Community/Junior College
XXXXXXXXXXXXXXXXXXXXXXXXXXXXXXXXXXXXXXXXXXXXXXXXXXXXXXXXXXXXXXXX
                 Vocational-Technical School
               XXXXXXXXXXXXXXXXXXXXXXXXXXXXXXXXXXXXXXXXXXXXXXXX
Vocational-Industrial or Area Vocational School
XXXXXXXXXXXXXXXXXXXXXXXXXXXXXXXX
```

Although the community, technical, and junior colleges cover all occupational education levels, from the single skill occupations to the technician and para-professional occupations, the predominant emphasis is on the upper three categories -- the skilled craftsman, technical specialties, and engineering technicians.

Using the upper categories to establish a base line for developing both entrance and graduate standards, an acceptable definition of technical education is necessary to demonstrate the necessity of general education. The following definition serves the purpose of this discussion:

> Technical Education -- That body of knowledge organized in a planned sequence of classroom and laboratory experiences, usually at the postsecondary level, to prepare pupils for a cluster of job opportunities in a specialized field of technology. The program of instruction normally includes the study of the underlying sciences and supporting

> mathematics inherent in a technology, as well as methods, skills, materials, and processes commonly used and services performed in the technology. A planned sequence of study and extensive knowledge in a field of specialization is required in technical education, including competency in the basic communications skills and related general education. Technical education prepares for the occupational area between the skilled craftsman and the professional person such as the doctor, the engineer, and the scientist. (U. S. Office of Education, 1969, p. 85)

Most significant in the definition is the inclusion of the reference to "the study of the underlying sciences and supporting mathematics inherent in a technology" as well as "competency in the basic communications skills and related general education". In stating the need of supporting mathematics, it is clear that technical education includes mathematics instruction. Technical courses and math courses are inseparable units in any technical education curricula. Technical education cannot be offered without the essential contribution of mathematics instruction. The relationship is symbiotic.

One hundred-eight senior executives from a random sample of Fortune 1300 companies endorsed the importance of this significant relationship. In a national survey of the firms, 76 percent responded that competency tests in writing skills and math should be administered prior to hiring of secretaries, clerks, first-line managers, and middle-level managers. Thirty-eight percent indicated that tests should be administered to hourly factory workers. Ninety-two percent also responded that public schools should be responsible for teaching computer use, with 58 percent indicating that such courses should be required and 42 percent indicating they should be optional (Research & Forecasts, Inc.).

Further, Koltai reports that " . . . the abilities of reading, writing, computing, and communicating are prized by business and industry above more specific job-related skills" (p. 21). There are those who counter this position by observing that if the job-related skills -- that is, the vocational/technical skills -- had not been developed at the time of hire, the general education skills of reading, writing, computing, and communicating would most surely move to a lower priority than the job-related abilities. This strongly suggests that a balance between the general education and occupational skills must be created to produce program graduates whose skills meet employer requirements and whose performance on the job substantiate a college's reputation.

The issues of integrity and consistency were foremost in the discussions of roundtable participants convened by the American Association of Community and Junior Colleges to formulate a policy statement on the meaning of the associate degree. The associate degree was reaffirmed "as central to the mission of the community, technical, and junior college" (p. 2). Explicit in this reaffirmation is the responsibility of the institution to "establish and maintain excellence in all educational programs . . . by providing the individual with the currency to negotiate the next step, whether that step be into full-time employment or into a baccalaureate degree program" (p. 2). Further, the organization of the curriculum "must include the opportunity for the student to demonstrate proficiency in the use of language and computation, for whatever their career goals, students will be called upon to exercise competence in those areas." The program should include "a level of general education that enables the individual to understand and appreciate his/her culture and environ ment" and "the attainment of skills in analysis, communication, quantification, and synthesis necessary for further growth as a lifespan learner and a productive member of society" (p. 3).

It is apparent in the educational process "that the buck stops" at the faculty level in resolving, on one hand, the commitment to standards by the institutions, and, on the other hand, the expectations by those receiving the graduates -- society in general, baccalaureate institutions, and employers. Standard-setting demands interaction among the program constituents -- faculty (both program faculty and general educatio faculty) and employers. The standards thus set should appropriately reflect the realistic competencies for initial employment as well as for career mobility -- employment adaptability skills to provide for upgrading and promotion. The success or failure of students should be attributed equally to the general education faculty and to the faculty in the student's major technical department. This "ownership" component will ensure agreement and appropriateness between the program content and the student's competency as certified by the award -- whether an individual course, a series of courses, or program completion.

The program specifications and expected outcomes must give consideration to the resources available for instruction over the time that a student plans to pursue his/her goal, and to the admissions standards to insure that the instructional program can facilitate the student's achievement of the stated competencies.

EVALUATION OF INCOMING STUDENTS

One of our most pressing problems in higher education is that students enter colleges with serious academic skill deficiencies. These deficiencies force institutions to provide remediation programs. A November 1983 conference called by the Conference Board of the Mathematical Sciences identified contributing factors in the current problem of math deficiencies that necessitate remediation.

Illustrative factors listed were:

1. The common practice during the past decade of "social promotion" of students in elementary and secondary schools.
2. Open admissions policies in public higher education.
3. The increased mathematical sophistication of work in many disciplines and professional training programs, leading to a rapid escalation of mathematics course requirements in high school.
4. Lack of societal support for mathematics education.
5. Lack of interest, motivation, and commitment on the part of students and faculty. (p. 16)

Developmental studies and remediation courses have as many working definitions as there are institutions and instructors in those institutions. For example, demonstrating the wide divergence of the definition of remedial mathematics is one offered in 1965 by Dr. Bruce E. Meserve, who was then serving as president of the National Council of Teachers of Mathematics, to a conference sponsored by the College Section of the American Textbook Publishers Institute held in cooperation with the American Association of Junior Colleges. He defined "remedial mathematics" as any course "which precedes calculus. College algebra, trigonometry and analytical geometry are some examples."

A review of the course listings of the nation's two- and four-year institutions would most likely reveal that a great majorit of mathematics course offerings, using Dr. Meserve's definitior are at the "remedial level".

A recent Washington Post article quoted University of Texas professor John E. Rouche, author of a report on remedial programs scheduled for release in the fall, 1984, as stating that "the colleges are caught between a rock and a hard place" . . . . and "they're stuck with students who want a college experience but lack the skills to handle it. On the other hand, their (the colleges') income depends upon enrollment so they've been forced to compromise." In the same article, Professor David Reisman of Harvard stated that "obviously, there have to be some standards to have achievement."

Universal access to college seems fair; but it appears that it has been destructive in that the level of achievement has slowly declined to meet the abilities of the learners. With 42 percent of national math enrollments in two-year colle falling in the remedial category, it is apparent that existing standards or those that are established must drive the develop- ment of any testing programs which may be used for screening. Only by using testing programs that are in agreement with the expectations of constituency groups will the results be satis- factory to the community as a whole and only then will the specification of remediation be effective and acceptable. Rouche's national survey indicates that only a third of the students pass remedial courses the first time that they take them and the eventual pass rate is uncertain.

Those in postsecondary occupational education would do well to endorse several of The Conference Board of the Mathe- matical Sciences' recommendations relating to remediation. Those recommendations include:

1. Identification of exemplary and cost-effective programs for remedial education.
2. Development of appropriate text and non-text materials for remedial courses.
3. Improvement of diagnostic testing techniques that will better identify student difficulties and lead to more specific, and perhaps, less costly remedial efforts. (p. 17)

Basic to the reduction and/or elimination of the remediation problem is the necessary progressive development of competency based materials for mathematics in grades k-12 to guarantee ability plateaus commensurate with the achieved grade level of the student. Evaluating students entering post-secondary institutions will then become more program specific, assessing the mathematical base needed for success in their occupational field rather than testing to determine if entering students are at the threshhold of functional mathematics.

The successful testing program must incorporate the thinking of the general and technical education faculties and employer representatives. In particular, the involvement of the occupational education faculty and the employer advisory groups will remove the threat commonly voiced: "The academic (mathematics) faculty keep our students from getting into the institution by setting high standards and, then, once in, prevent them from completing because of the unrealistic expectations and the irrelevance of the content to our program."

In effect, sense can be made of the situation only through cooperative endeavors.

Ownership makes the difference.

SPECIFICATION OF REMEDIATION

The equation, (HS + RE) + PS = AAS/DIPLOMA/CERTIFICATE, perhaps, quantifies conceptually the situation in which we find ourselves. The equation symbols are: HS = high school mathematics; RE = remediation: PS = postsecondary mathematics. Mathematics instructors will agree most likely that each of the elements of the equation must either be known or have a finite value to result in a solution. However, because the equation does not represent quantifiable definitions, it is subject to the frailities of human nature. (HS + RE) + PS = AAS/DIPLOMA/CERTIFICATE is very appropriate to our application except when the expectation of RE exceeds the resources or the specification of RE is less than appropriate for the problem. However, the long range goal should be to reduce the equation to HS + PS = AAS/DIPLOMA/CERTIFICATE.

Society's expectation of high school graduates is that they will be prepared for either the world of work or for higher education. Curriculum developers in two-year institutions design programs based upon their knowledge of the skills high school graduates are certified to have and upon the assumption that students will bring twelfth grade competencies to the institution. The remediation effort being expended nationally by higher education is evidence that our expectation and assumptions are faulty. Therefore, to produce a graduate who meets the expectations of the institution and the program major, remedial education has been introduced to "guarantee" that our graduate will achieve the stated standards.

To meet this "guarantee", the burden of bringing the student "up to speed" is usually shifted from the occupational education department to the mathematics department and, often, from the mathematics department, to the remediation or developmental studies department. Here, too, there is a need for partnership. The remediation program too frequently resembles high school mathematics programs. Most frequently, little or no relationship exists between the college occupational or mathematics faculties and the remediation department. These departments too often abrogate their responsibilities and await the results of the remediation only to find out too late that the diagnostic and prescriptive procedures were faulty and, as a result, the deficiencies still exist to a degree great enough to inhibit the student's full development and graduation. Without partnership and the active involvement of the prime investors -- occupational, mathematics, and remedial faculties -- the remediation will fall short and any deficiency will be the fault of the other party.

Ownership is the difference.

MAINTENANCE OF STANDARDS

The academic integrity of any institution is vested in the several constituent divisions -- administration as it relates to leadership, admissions as it relates to selectivity instruction as it relates to learning, student welfare as it relates to counseling. All over the country in graduation exercises academic officers state to their presidents and to the audiences: "The faculty have reviewed each student's achievements and certify that these students have met all of their academic obligations and program requirements for the awards they are about to receive." This certification symbolizes the "academic integrity" of the institutions that is vested individually and collectively in the faculty and all college divisions.

Graduation exercises highlight an important conflict that often exists between general education faculty and their standards. The success or failure of graduates to demonstrate competencies gained from general education courses is most often not measured by competency on the job. The measures of quality all too often are the statistics of grade distribution. If too low a distribution is observed, the tendency is to be concerned with the admissions process. This initiates a feedback process which often results in the review of admissions standards and, may yield higher standards for admission or prerequisites for entrance into the regular general education courses.

The reaction of the occupational faculty when general education standards are set arbitrarily is to review the areas of course weakness -- often taken as those courses which "weed out" occupational students and prevent them from gaining an employable skill -- and recommend the elimination of certain courses from the curriculum on the basis of irrelevance. Or, more often than not, the occupational faculty insist that the general education courses are "just taking too much of the credit load for our students and the mathematics and physical science, etc., can be incorporated into the major courses without loss of content."

And, then, another integration occurs which places a general education course within the occupational department to "guarantee" quality control and graduation.

CONCLUSION

None of the problems that I have sketched in this brief presentation are insurmountable. All of them are significant, however, and they must be addressed if our colleges are to maintain their integrity, if they are to continue to graduate students who have competencies valued by employers, and if the colleges are going to continue to help skill-deficient students to develop to a level adequate to successfully complete technician level study programs. These problems cannot be addressed only by administrative fiat, although administrative mechanisms should be applied when necessary.

The central players in this effort, however, are the faculties of the general education and occupational departments along with employer representatives. Their relationship must be a partnership, with each contributing equally to the work and with each sharing the accolades of success -- and the blame for whatever failures occur.

The concept and characteristics of ownership must be developed and accepted by general education departments/faculty so that a failure of a student or graduate is as impressed upon the general education faculty as it is upon the vocational/technical faculty. Only then will the establishment of standards, evaluation of incoming students, specification of remediation and the maintenance of standards serve the institution and students in a responsible fashion. Each partner must feel "ownership" in the program with a vested interest in the graduate.

Ownership is the difference.

Ownership makes the difference.

REFERENCE LIST

American Association of Community and Junior Colleges. 1983 technical and junior college directory. Washington, D.C.: The Association, 1983.

American Association of Community and Junior Colleges. National task force to redefine the associate degree: a preliminary presentation. Washington, D.C.: The Association, April 1983.

American Association of Community and Junior Colleges. Policy statement on the associate degree. Washington, D.C.: The Association, March 1984.

American Textbook Publishers Institute, College Section. The teaching of the remedial student in English and math (a conference report held in cooperation with the American Association of Junior Colleges). Reprinted from Publishers' Weekly, January 3, 1966.

The Conference Board of the Mathematical Sciences. New goals for mathematical sciences education (a conference report). November 1983.

Greer, J. S. The organization and administration of technical education for industrial occupations with emphasis on teacher education (Doctoral dissertation, University of Connecticut, 1967). Dissertation Abstracts International, 1967, 28, 8-A. (University Microfilms No. 68-01350).

The National Commission on Excellence in Education. A nation at risk: the imperative for educational reform. Washington, D.C.: U. S. Government Printing Office, April 1983.

Research & Forecasts, Inc. Business Poll: U. S. education system is major concern for business. New York: Research & Forecasts, Inc., Fall 1983.

U. S. Office of Education. Vocational education and occupations. (USOE Publication No. OE-80061). Washington, D.C.: U. S. Government Printing Office, July 1969.

Vaughan, G. B. Historical perspective: president Truman endorsed community college manifesto. Community and Junior College Journal, 1983, 53(7), 21-24.

The Washington Post. Remedial work seen as erosion of education. April 29, 1984.

(DISCUSSION BEGINS ON P. 131.)

TECHNICAL MATHEMATICS

IN TWO-YEAR COLLEGE PROGRAMS

by

Allyn Washington

SUMMARY

Technical and vocational mathematics courses are service courses which cover the mathematics, and appropriate applications, needed in technical and vocational programs. The development of the mathematics is non-rigorous. The applications show where and how mathematics is used, but do not require a knowledge of a specific technical subject by either the student or the instructor. A primary reason for the establishment of these courses is that traditional mathematics programs do not meet the needs of the technical programs in material development, timing of topic coverage and applications. Motivation of students, who are often not well prepared academically, is a major reason for the emphasis on applications.

Technical mathematics courses should be coordinated with the technical courses of the program. The best way to accomplish this is through formal and informal meetings with technical faculty.

The calculator is used in all technical mathematics courses and is a great assist in teaching many topics. The computer is also used, although it is generally not integrated directly into the course as is the calculator.

Many mathematics faculty persons prefer not to teach technical mathematics because of the applications, and many do not prefer to teach vocational mathematics because of the mathematical level. This problem has no easy solution, although special in-service training could provide the background in applications that instructors would like to have.

Technical and vocational mathematics courses will continue to grow due to the strong emphasis on technology in industry.

TECHNICAL MATHEMATICS IN TWO-YEAR COLLEGE PROGRAMS

Allyn J. Washington

I. The Development of Technical Mathematics

Mathematics is the basic tool used by technicians of all fields to analyze and solve problems which arise in applied situations. This means that technicians require a knowledge of mathematics, and that technical programs should include appropriate mathematics courses.

There have been many college mathematics courses developed which have been directed specifically at students enrolled in various vocational and technical programs. These courses are now generally labeled as vocational, industrial or technical mathematics. The purpose of these courses is to provide the mathematics necessary in the curriculum, with a greater emphasis on identifying and using applied problems which are appropriate to the students' program. The mathematics topics covered, the level at which the courses are presented, and the emphasis given to the applications vary considerably with the type of program, the college, and the instructor.

Technical and vocational mathematics courses are primarily service courses. The emphasis should be to develop properly the mathematics which is needed and used in other program courses, including mathematics courses. The development of the material is done in a non-rigorous way. Applied problems are an integral part of such courses, and the applications should be directed at the students' curriculum to the greatest extent possible. However, the use of applications serves to show where and how mathematics is used and does not require prior knowledge of a specific technical subject by either the student or the instructor

The American Mathematics Association of Two-Year Colleges (AMATYC) has established a standing committee on technical mathematics. In the proposal (1) for establishing the AMATYC Technical Mathematics Committee the "newness of the technical mathematics movement" was noted. This refers to the rapidly growing number of courses at two-year colleges and the general awareness today of such courses under the general heading of technical mathematics. However, it should be noted that some textbooks which are in use today are revised editions of texts which were used in courses especially designed for certain technical programs as early as the 1920's. Also, a number of textbooks intended for use by students in technical programs were developed during the 1950's and 1960's. Sessions dealing with technical mathematics were held at statewide (New York) faculty meetings in the 1950's.

There are many reasons which have led to the development of technical mathematics. The proposal for the establishment of the AMATYC Technical Mathematics Committee states that "traditional mathematics programs are not meeting the needs of these technical programs and most institutions are finding the need to offer two different tracks of mathematics education." The traditional programs usually do not have the less formal material development, timing (related to the needs of the other technical program courses) of topic coverage, and application coverage which is appropriate for technical students and programs. About ten years ago I made a study of over 100 two-year college catalogs and found that about 75% of the colleges offered mathematics courses specifically designed for technical students, while about 10% had their technical students take the traditional types of courses with the students of other programs. The remaining 15% did not offer these types of technical programs. Although my survey has not been updated, the Conference Board of Mathematical Sciences survey (2) in 1980-81 reports that "many two-year colleges ... have greatly expanded their scope to include a host of vocational programs." One of the highlights of the report was that "in two-year colleges occupational/technical program enrollments now lead college transfer enrollments."

II. Applications and Technical Mathematics Courses

Helping to motivate students enrolled in technical and vocational programs is one of the major reasons for the emphasis on applications. When the mathematics is related to their curriculum through the use of appropriate allied problems (rather than the approach where applications are used infrequently), students better understand the need for the mathematics that they are expected to learn.

Many mathematics educators have observed that the use of applications does help motivate students in the learning of mathematics. William A. Granville in the preface to his 1911 calculus book (3) noted that "simple practical problems have been added throughout; problems that illustrate the theory and at the same time are of interest to the student." Clarence E. Tuites in the preface to his 1946 technical mathematics text (4) says "representative problems from the various technical fields have been selected so that a student with a particular interest in one of these fields may be given an assignment of practice problems illustrative of the use of mathematics in his chosen field." The CUPM Panel (5) in 1982, in giving their recommendations for minimal mathematical competencies for college graduates, stated that courses "should be designed to be appealing and significant to the student", and that "almost all undergraduate courses in mathematics should give attention to applications." A study (6) at Stanford University in 1983 reported that providing general motivation in the class was found to be an important point in making a class easy to follow. The report states that "usually the motivation has to do with the expected future use, with repeated applications, or with other factors of relevance to the students". Also from my own experience of over twenty years of teaching technical mathematics I can say that it has been many years since I heard the question "What is this good for ?"

One point should be reemphasized. Although applications are an important part of a technical mathematics course, the primary content is still mathematics. Frequently it is the only mathematics course which a technical student will ever take in college, and it should provide a good mathematics background. The student may wish to take more mathematics at some time in the future, and most employers of technical program graduates want students to have had a good general background so that they may be more adaptable to a variety of possible job opportunites. Over a period of years in meetings of technical program advisory committees and in many personal conversations with people from industry, I have heard employers ask for certain specific skills and a good general background. Also, an informal study (7) reported in the American Mathematical Monthly that employers of graduates from four-year mathematics programs also prefer a good general background as well as specialized courses.

To some, the meaning of technical mathematics is more general than to others. In its more general meaning it includes nearly all mathematics courses which are directed at specific two-year career programs, as opposed to the courses which are intended for a general student audience or for students intending to transfer to four-year colleges. In its more restricted meaning it includes those courses which are directed primarily at engineering related technical programs such as electronics technology, mechanical technology, chemical technology and architectural technology. The courses intended for rather specific career programs, and which generally require a somewhat lesser level of mathematics, are often referred to as vocational, industrial or occupational mathematics. These would include courses for students in technologies such as automotive, construction, electricity, air-conditioning, machine, drafting, police science, agriculture, fire science, forestry, and welding. Courses for other career programs such as business technology, data processing and nursing are sometimes referred to as technical mathematics, although they are also commonly labled business mathematics or some similar appropriate title. (In this paper the more restricted meanings of technical and vocational mathematics are intended.)

Vocational and industrial mathematics courses are often taught by the faculty of the vocational programs, rather than by faculty of the mathematics department. Such courses often emphasize applications to a much greater extent than mathematical content. Technical mathemetics courses are generally taught by faculty of the mathematics department. Often these courses do not emphasize applications to the extent which would be appropriate for the students' programs. The mix of mathematics and applications which is appropriate for the course in any given program should vary with the needs of the program and the students who are enrolled. Designing the course properly requires a great deal of planning by the department as well as the instructor.

Technical and vocational mathematics courses should be coordinated with other program courses so that the mathematics topics are covered before they are encountered in these other courses. To do this requires consultation and coordination with the technical departments. Such consultation may reveal, for example, that electronics students need determinants and trigonometric curves at certain times, and that mechanical students do not need them, whereas the reverse may be true for conic section curves. Thus, courses must be developed which do not simply follow a traditional sequence of mathematical topics and which develop course content that is not simply a review of material which the student may have had previously.

The mathematical content of technical mathematics courses generally consists of algebra and trigonometry. Some technical programs such as electronics technology and mechanical technology often require some of the more advanced topics in algebra and trigonometry as well as some calculus. Since the courses should be designed to be coordinated with other technical program courses and are often trying to upgrade a student's skills, such courses very often integrate coverage of topics in algebra and trigonometry. With the applications and topics from algebra, trigonometry and calculus, a great variety of course sequences is possible, and in fact is found in two-year college programs. Another feature of a well designed course is that it should proceed somewhat slower at the beginning so that students can succeed and realize that the material can be learned.

A technical mathematics course covers most of the basic topics in algebra and trigonometry, although some of the more advanced topics (e.g. radical equations, matrices, trigonometric equations) are frequently not covered. In the basic coverage, only a few topics such as the more advanced methods of factoring, or the sum and product of the roots of a quadratic equation, are generally not covered. Besides applications, graphing is used more extensively. It is covered whenever it is appropriate (e.g. when discussing linear equations, quadratic equations, trigonometric functions, equation solving, and so on.)

III. The Technical-Vocational Mathematics Student

Students enrolled in technical and vocational programs are usually not as well prepared academically as students intending to major in such programs as mathematics, science or engineering. Some of these students have well-founded career plans when they enter college, whereas many choose technical or vocational programs not long before (or after) entering. This latter group includes those who have chosen the program because it will lead to a career "where the jobs are" (e.g. electronics or computers) even though they really had not planned on or are truly committed to such a career. Also, many of these students were not serious students in secondary school. Therefore, motivating this group of students in much of their academic work in college becomes a major undertaking.

The mathematical needs of the students in vocational and technical programs depend on the backgrounds of the students and the program in which they are enrolled. A great many students enter vocational programs with weak backgrounds, in terms of both their mathematical skills and their attitudes towards learning. Although vocational mathematics courses generally consist of topics from arithmetic, geometry, basic algebra and trigonometry and appropriate applications, many students have difficulty in handling these topics. It is necessary to develop these topics in such a way that they are not handled as strictly review in nature. The applications which are used provide a means of introducing mathematical topics as well as motivating students by showing where and how the mathematics is relevant to their program. The instructor must work closely with the students in showing them that they can indeed learn and use mathematics.

Students in technical mathematics courses generally have taken more mathematics than vocational program students in secondary school. Their backgrounds generally include elementary algebra and geometry, and often intermediate algebra and some trigonometry. However, they often do not know the material well and are unaware of this. Again, applications can be used to introduce mathematical topics as well as to show the relevance of the mathematics being covered.

IV. The Calculator and the Computer

The calculator is now used extensively throughout technical mathematics courses. It is also used in vocational courses, although when arithmetic is the topic its use is often restricted. In most technical mathematics courses the students are expected to have a scientific calculator, and most students are able to use it with a minimum of guidance. The calculator is of great value in teaching trigonometry. My experience (and that of many others) shows that topics such as triangle solution and vectors are now learned with much more success than in pre-calculator days. Students now find that they can get the right answer, and therefore understand the methods. Previously, they frequently could not handle the slide rule, came up with the wrong answer, and thereby thought that they did not understand the methods involved. There is little doubt that the calculator is of great assistance to students in learning and understanding technical mathematics.

The computer is also used in many technical mathematics courses. The computer language which is used more than any other is BASIC. However, the use of the computer is not often integrated directly into the course in the same way as the calculator. Computer programming and specifically computer related mathematical topics are generally not included. (These conclusions are found in a summary of textbook reviews which I studied in 1983.) Instructors use the computer to solve certain types of problems, but generally the students who use it are familiar with the computer through other courses. However, there is no doubt that the use of the computer is increasing rapidly.

V. Technical-Vocational Mathematics Faculty

It has been frequently noted that applications are an integral part of a technical or vocational mathematics course. However, some mathematics instructors are hesitant to include many types of applications when teaching these courses since they are not very familiar with them and the related technical and vocational areas. This can be a difficult obstacle to overcome, and is one of the more serious problems found in the teaching of these courses. The instructor must realize that the applications show where and how the mathematics can be used, but that he or she is not expected to be expert in all technical areas. In fact, this can give the instructor an excellent opportunity to involve the students. The instructor, in stating that he or she is not very familiar with a certain applied situation, will often find that the students are eager to show the knowledge which they have gained from other technical program courses.

Many mathematics faculty persons do not prefer to teach technical mathematics because of the applications, and many do not prefer to teach vocational mathematics because they feel tha the mathematical level of such courses is lower than they enjoy teaching. Getting these faculty members interested in teaching these courses is difficult. Involvement with the computer and seeing its applications has led some to find that such courses can actually be quite exciting. Becoming well acquainted with faculty in the technical areas, sitting in on an occasional technical course class or laboratory session, taking technical courses in retraining programs or on sabbatical leave also shows some that these can be interesting courses. However, unless there is motivation on the part of these instructors, they tend to keep their reservations about teaching these courses.

When developing or updating technical and vocational mathematics courses, it is necessary to be aware of the needs of the program being serviced. Attending sessions at conferences which are devoted to technical and vocational mathematics and being aware of the textbooks which are available can assist in this. However, one of the very best ways is to have meetings, formal and informal, with the members of the technical departments of the college. This provides the input and communication necessary to develop very useful technical and vocational mathematics courses. It also gives an excellent opportunity to interest mathematics faculty in technical programs. Another obvious benefit is that it can lead to a much better working relationship between departments.

The statements and conclusions just made regarding the preferences of technical mathematics instructors and the need for close consultation with technical departments are made after years of personal observations and numerous conferences and informal contacts with faculty in mathematics and technical departments. However, they are strongly reinforced by a study by R. M. Davis in 1980 (8) of service courses at two-year colleges. Davis states that "mathematics faculty have limited interest in teaching occupational-technical mathematics courses", and that the "level of interface (between departments) is a key to successful occupational-technical programs."

The updating and upgrading of faculty is a general problem. Some take courses during summers and evenings; a few manage to get sabbatical leaves for professional development; some attend special workshops or conferences. Among the problems are scheduling of classes to allow time to take courses, difficulty in obtaining sabbatical leaves and financial reimbursement, and conflicting personal plans. Technical mathematics faculty probably need updating in two principal areas. One is computers, although the problem here appears to be dissipating rather quickly. The trend to use computers is very strong, and most instructors are now familiar with them. The other is, again, in the area of applications. The reluctance of instructors to use and assign applied problems can often be overcome if they become knowledgable in one or two areas. Then they generally are willing to use the applications in other areas as well, even if they are not very familiar with the specific types of applied problems. Special training programs such as summer institutes, specific short-course programs, or in-service work in the private sector would readily provide the background in applications. Such special training programs should be established.

VI. A Look Ahead

Looking ahead, it appears that technical and vocational mathematics courses, along with nearly all mathematics courses, will continue to grow in number and in acceptance. One of the major reasons for this is the extremely strong emphasis on high technoloy in industry today. Technical program graduates in the years to come will need even stronger academic backgrounds, and mathematics must be one of the major components.

(1) Proposal For Standing Committee of AMATYC, Technical Mathematics Committee (March 31, 1981)

(2) Conference Board of the Mathematical Sciences, Report of Survey Committee, Vol. VI, 1980-81

(3) W. A. Granville, Elements of the Differential and Integral Calculus (Revised Edition), Ginn & Co., 1911

(4) C. E. Tuites, Basic Mathematics for Technical Courses, Prentice-Hall, 1946

(5) CUPM Panel, Minimal Mathematical Competencies for College Graduates, American Mathematical Monthly, April 1982, pp. 266-27

(6) N. Hativa, What Makes Mathematics Lessons Easy to Follow, Understand, and Remember ?, Two-Year College Mathematics Journal, November 1983, pp. 398-406

(7) A. K. Stenney, Undergraduate Training for Industrial Careers, American Mathematical Monthly, August-September 1983, pp. 478-481

(8) R. M. Davis. The Development and Delivery of Mathematics Service Courses at Two Year Colleges, Ph.D. dissertation, U. of Maryland, 1980

(DISCUSSION BEGINS ON P. 131.)

DISCUSSION

TECHNICAL MATHEMATICS

1. A New Approach to Remediation

2. Who Should Teach Technical Mathematics?

3. Threats to Technical Mathematics

4. F.O.W.G.I.

A NEW APPROACH TO REMEDIATION

Leitzel: What would be the effect of taking students who generally would take basic skills courses in arithmetic and algebra and starting them instead in the beginning technical mathematics sequence?

Rodi: I think that remedial courses are, in part, looking to the next course. Very explicitly, you are preparing somebody for Algebra I or Algebra II, or whatever. Tech math courses have as their more immediate goal to learn a topic because it's going to be needed in a drafting course for a practical application. Tech courses in a certain sense are terminal in themselves mathematically. They're preparing students for something in the tech area; remedial courses I am assuming are not terminal in themselves--they're aiming at some other math course further down the curriculum.

Albers: I'm hearing a commercial in a way for doing something with remediation along with very strong applications. I don't think existing remedial books have many applications.

Rodi: In response to the initial question: "Could we put remedial students in those particular courses?", I happen to think you probably could.

Davis: Yes, you could put remedial students in there, but you as a teacher would come to a point and say, "Do I give up a little of the math to make sure the students comprehend how what they have learned fits into the technical area?" Or, "Do I give up some of those applications and some of those explanations so that I am sure they will have all of the math skills, the bag of math tools for their next class?"

<u>Who Should Teach Technical Mathematics?</u>

Page: If you are going to have tech applications, then you need somebody who is a specialist who can teach them; whereas, if you have a broader course which interfaces with various technologies, then the person who is teaching that course does not have to be a specialist in that particular technology but must be more broad-based.

Washington: In working with the technical people over twenty years, I have found that as long as you cover the math topics they needed, by the time they need them, it went very smoothly;

there was never any adverse reaction from them. It was really a matter of timing.

Rodi: There are not enough math people in community colleges who are really interested in teaching tech math because most of them have been trained in programs that really sought to prepare them to teach undergraduate mathematics; but looking at the long-range needs of the country over the next twenty-five years, we will need a lot of people who are very well-trained in technical mathematics, and I think one of the ways to address that problem is to start training such faculty.

Rodi: Tech math courses should be taught by the math department. In a way, this is the university problem repeated in a different form. Statistics courses are taught by many departments all over the university--because university math departments, I think, have not been willing to adjust what they are teaching to what the engineers want or what the psychologists want or whatever, and in many ways I think they've cut their own throats over the last twenty-five years by doing that; I feel very strongly that if the content is basically mathematics, then the math department ought to be teaching it. I think you have exactly the same problem at community colleges with technical math programs. The automotive people want to teach their math course, the drafting people want to teach their math course, and the accounting people want to teach their math course--we've had about four courses get away from us that way. I think it's as short-sighted in the community colleges as it is in universities for mathematicians not to realize that an integral part of mathematics is applications of mathematics.

Davis: The best training we could give people who are going to teach such a course is to send them off for a year to spend three months with a construction company, three months with an electronics company, and three months with an automotive company, etc. They would come back with a much better understanding of what is needed.

Leitzel: I think there are ingredients in the industrial courses that would serve to greatly strengthen the basic skills instruction that we are doing. We do know that students with weaknesses in mathematics identify with concrete problems and can solve concrete problems. They can be helped by approaching mathematics through those kinds of problems.

Perhaps we haven't given enough attention to learning from those courses what the appropriate aspects are for our other instruction.

Rodi: There is another reason why these tech courses ought to be taught by math departments. While they keep the basic nature of technical mathematics, students are going to get a little more of the abstract structure that is going to be applicable in a variety of situations when they are taught by math departments. I'm afraid if these courses were taught out of the technical area there would be a little more tendency to orient them to "just learn how to do this" and not worry about how the general structure might have to be modified ten years from now. And that's why I am sort of passionate about math departments not giving them up.

<u>Threats to Technical Mathematics</u>

Warren: At graduation, one of the dichotomies that I see as I stand in the foyer of our auditorium as our graduates come in is that all of the tech faculty are there wishing the graduates well, shaking their hands, but the general education faculty are absent except for a few. I don't believe math faculty have the "ownership" feeling as do the technical faculty--maybe for a very good reason. Our graduates go out and the first complaint we hear from employers, if we hear a complaint, is that they are not technically competent. Employers don't call up and say that they can't write well, they can't compute, they can't use a calculator or a microcomputer, or that they can't provide the calculations. I think the greatest threat to good quality vocational/ technical education across the nation exists in the Joint Training Partnership Act, the child of CETA. There are many employer groups determining what educational programs are going to be. They come to my institution and want us to turn out electro-mechanical technicians in a six-month period. Now you can't turn out electro-mechanical technicians in a six-month period--you can barely turn them out in a two-year period. But yet those employers are looking at the specific skills that they need--the specific little piece of mathematics, physical science, strength of materials, and chemistry that they need and say to us "teach that and that's all." You can get students into their industries, but then they lay them off in six months and send them back to us for retraining. They are no longer employable because they have no transferable skills.

Page: Local industries have a vested interest also in technical education. I don't see the colleges tapping into this vested interest, and I don't see the faculty willing to share that ownership with a broad enough partnership. I don't see industry contributing economically, politically, or in other ways. Now, you do allude of course to the fact that we do have advisory commissions, but I think it's in industry's own interest to provide more than that--to provide scholarships, exchange programs, and training programs.

FOWGI

Warren: You have to use the FOWGI--"Find Out Who's Got It"--approach. I think you have to go to industry and ask. As a prime example, we have a one-year plant and soil technology program at my campus which is 100% transferable to the University of Maine. We had one of the industrial advisory committees come up and provide us with 100,000 dollars for a one-faculty/twenty-student program because they knew what our needs were. Proprietary schools are taking over much of specialized education, and they are going out and writing contracts with IBM, Intel, and other companies. The companies are picking up the entire tab plus buying into the institution to take care of administration.

Leitzel: What are the possibilities of linking two-year programs and business and industry for the employment of faculty? I'm thinking of math faculty who desire more technical experience. The employment could be in the summer, for six months, or for a calendar year for the purpose of upgrading the faculty, broadening their experience, and giving them more familiarity with the needs of industry.

Warren: I think there's a great opportunity. I have eight faculty in my electrical/electronics department, and probably five to six of them go into industry every summer and spend eight weeks working for major paper companies or electro-mechanical companies while learning what has happened in the last twelve months.

Part 3.

ENDANGERED CURRICULUM ELEMENTS

REFLECTIONS ON BASIC MATHEMATICS PROGRAMS IN THE TWO-YEAR COLLEGE

by

Geoffrey R. Akst

SUMMARY

Each of the sections of this paper develops the theme of basic math programs in the two-year college (TYC) from a different perspective. The first section recalls some of the factors which contributed to the growth of these programs in the 1960's and 70's. The second addresses the question how well basic math programs, as they have developed, actually work. The third contains suggestions for making these programs work even better. And finally, the fourth section deals with strategies for self-destruction, that is, for changing the context in which we operate so that one day TYC basic math programs will no longer be necessary.

REFLECTIONS ON BASIC MATH PROGRAMS IN THE TWO-YEAR COLLEGE

Geoffrey Akst

The Growing Need.

Not only is the problem of mathematics remediation in the two-year college (TYC) serious, but it is getting worse. National statistics tell the sad tale. In 1980 (the most recent year for which information is available), courses in basic mathematics -- arithmetic, elementary algebra, intermediate algebra and general mathematics -- accounted for nearly half the total TYC math enrollment, a good one-and-one-half times what the corresponding proportion had been in 1966.[1] Furthermore from 1966 to 1980, it was the group of very weakest students that grew most rapidly; as the TYC movement expanded, with their total math enrollment tripled, the registration in arithmetic exploded, zooming to eight times its former size. What is so disconcerting about this development is that arithmetic was the only math that many of these students seem to have studied in high school! For every TYC student who took calculus in 1980, there were three others taking arithmetic; for each student taking elementary statistics six were learning (or relearning) computational skills. In all, close to half a million TYC students in 1980 were registered in courses covering secondary or even primary school mathematics.

This inadequate preparation of incoming students is not a problem confined to TYC's or to mathematics. Verbal skills of entering TYC students have declined as well [2], and four-year institutions, even the most selective, have also begun to complain about the academic preparation of entering freshmen.[3] How are we to explain this dislocation in our educational system? The non-selective entrance criteria of the proliferating TYC's and of other open-admissions institutions was, of course, an aggravating factor. Relaxing college admissions standards weakened the incentive for high school students to apply themselves, and many grew confident of finding a niche in college no matter what their level of preparation. It is hardly surprising then that since the late 60's, the proportion of students following the preparatory track dropped sharply, and the less stringent "general" track became dominant.[4] Curriculum makers in the high school, under the pressure of student activism and liberated by diffuse college standards, allowed students to take fewer demanding courses such as algebra and geometry and to substitute electives like "personal development" courses. Across the curriculum, grade inflation and social promotion also contributed to declining expectations. Furthermore, an analysis of high school transcripts from around the country suggests that this phenomenon was more severe in mathematics than in any other discipline.[4]

As standards slipped in the 60's and 70's, the percentage

of students whom the high schools graduated increased, climbing from 60% to 75%.[5] More and more students were in a position to go on to college, and in those two decades, the college-age population actually attending college rose from 1/5 to 1/3.[6] Of course, many of the weaker high school graduates headed straight for the open-admissions institutions -- not only students who had failed to develop their capabilities but others of low academic potential who would not have been able to pursue a degree under earlier stringent conditions even had they applied themselves.

There may have been other less tangible societal pressures behind the slipping skills of incoming college students. Many observers noted signs of slackening discipline in the home and on the street which could well have impinged on student motivation and performance. A generation of television babies was growing up, many of whom were weaned less on Robinson Crusoe than on Gilligan's Island; the tradition of spending time interacting with books, conversing, and doing homework seems to have been in decline. Outside the home, heightened political unrest channelled the time and energy of many students away from academic pursuits. But the two factors first noted -- the reduction of entrance requirements at the postsecondary level and the weakening of secondary school exit requirements -- were most directly responsible for today's huge remedial enrollment. This causal relationship is important to keep in mind if the process is ever to be reversed.

How Well is the Need Being Met?

Given that TYC basic math programs have grown into sizable enterprises requiring the allocation of hotly contested college resources, the question is, do they work? The answer is an unequivocal Yes!, Maybe!, and No! What makes the evaluation of these programs so subjective is not just a matter of having to decide whether a glass of water is half-filled or half-empty; the more fundamental issue is which glass to look at. The problem is that the perception of the goals and expectations of a basic math program depends in large measure on the viewer's responsibilities and interests.

At one extreme, the faculty in the trenches -- those teaching basic mathematics -- will likely feel that their work is successful to the extent that students learn the remedial topics thoroughly and then pass their course. By contrast, instructors of client courses -- courses which cover freshman math, science, technology, health, business, etc. -- will be more concerned with the retention and carry over of the remedial content. Of course, the two groups of faculty may overlap to some extent, and it is amusing to see how my own feelings change as I shift roles from arithmetic to statistics teacher, condemning the low standards I employed earlier in giving a marginal student the benefit of the doubt. There is also the view from the seventh floor, to use the slang at my institution. Deans and presidents consider special programs such as mathematics remediation to be interventionist strategies; from their point of view, these programs succeed or fail depending on the extent to which they increase the number of students the college will retain or graduate.

All three perspectives are legitimate, and none can be ignored in any serious attempt to gauge program effectiveness.

From the remedial faculty point of view, measures of success in the basic math course tend to involve two types of data: final exam scores and final course grades. Happily, analysis of either can generally be relied on to yield positive results. Final exam scores, for example, virtually always turn out to be significantly higher than initial placement test scores, often by several standard deviations or more. Such large gains, however, may be deceptive. For one thing, design and statistical biases tend to inflate the gain beyond its true value; for another, genuine learning may be due to factors other than course participation.[7] As for course grades, between 1/2 and 3/4 of starting students in a typical basic math course pass.[8] Sometimes passing rates are significantly higher -- in summer session say, when the population is self-selected, or in sections with particularly effective instructors. Some basic skills courses will have lower passing rates, of course, because the students are unusually weak, the pedagogy is faulty, or too much material has been crammed into the course. It is important to bear in mind, however, that even when the passing rate is as high as 3/4 per course, the passing rate in the course sequence may still be discouraging, say 1/2 for arithmetic and elementary algebra, or 2/5 for the three-tier sequence from arithmetic to intermediate algebra. In fact, a recent survey of TYC's around the country found that about 40 to 50% of students starting basic math programs satisfy all their basic math obligations.[9]

How well do remediated students do in client courses? The best evidence is that about half of the students exiting from basic math programs go on to succeed (get C or above) in college-level math courses.[9] Is 50% good enough? Well, that figure is very likely lower than the success rate for exempted students, but higher than what it would have been had the remediated students not taken and passed basic math courses. Again, is the glass half-empty or half-full?

From the administrator's perspective, retention and graduation rates are key statistics in evaluating basic skills programs. At some TYC's, students who satisfy their basic math obligations are retained at a higher rate than those originally exempted [10] -- a respectable standard to meet; yet even here, the overall graduation rate may be disappointing, and one scans the faces at commencement spotting only an occasional student who started out in arithmetic. A recent study at CUNY's nine TYC's reports that only a third of all entering students eventually graduate -- only a fourth of those who come in with low high school averages. (By contrast, the graduation rate for CUNY's four-year college students, who generally enter with stronger academic skills and with more financial security, is a full half.[11]) One-third also seems to be a quick-and-dirty approximation for the TYC graduation rate on the national level as well; each year, TYC's around the country award only about 350,000 associate degrees [11] even though for each of the past ten years, they admitted one to one-and-a-half million first-time students.[6]

A pessimistic interpretation of such data is that basic skills programs as a whole are not compensating sufficiently for the deficiencies which incoming students bring to the TYC experience, and that the cost to the taxpayer, the TYC, and the student -- in terms of time, resources or money -- is simply unacceptable. The optimist will point out that TYC's prepare many of their non-graduating students for transfer to the four-year college of their choice, to advance on the job, to meet personal goals which may have nothing to do with graduating, or possibly to master enough skills which will enrich their lives and increase the contribution which they make to society.

Given that the picture is mixed, where does that leave us? The answer lies with some healthy realism. For one thing, to expect TYC's or basic skills programs to be responsible for ensuring that weak students overtake those who are stronger may not be reasonable. By seeking out the lowest achieving students, basic skills courses are in effect programming themselves for partial success at best. The main point is that TYC's, if they are to survive, must meet as well as they can the needs of the students they are able to attract. And rather than dwelling on what is probably an inherent limitation of the effectiveness of basic skills programs, it is more productive for those of us working in these programs to search for ways either to improve the services which we deliver to our students or to modify the conditions which originally contributed to mathematically handicapped students entering the TYC's.

Strategies for Meeting the Need.

It is to the first of these concerns that we now turn: how can TYC basic math programs be improved? At the risk of sounding overly prescriptive, I've distilled from my experience observing numerous programs a list of five broad guidelines which might serve as points of departure for increasing program effectiveness. Of course, there is more than one way to carve a turkey, and several readers may wish to meet me later to take issue with one point or another.

(1) The curriculum must be appropriate.

The curriculum should be chosen frugally, keeping in mind the program's propaedeutic function: preparation for specific followup courses, for passing a competency test, etc. Due to the pressure of time, non-essential topics must be kept to a minimum. Topics should be developed heuristically, with the acquisition of skills stressed. As to the calculator question, it is critical that arithmetic courses continue to cover conventional algorithms, but that they also teach the use of the calculator (estimation, constants, memory, scientific notation, etc.). This duality is essential for preparing students for situations at school or on the job which they are likely to encounter, some requiring the use of calculators and others forbidding it. Instruction and testing should emphasize applications and word problems, with many examples taken from followup courses, so as to promote carry over.

(2) Students who need help should get it.

A placement test should be given to all entering students, and used to identify which of them lack the competencies they need; a test has the advantage over the high school record of providing uniform, recent information on what students know. Sample questions from the exam should be sent to incoming freshmen before they are tested to encourage review. The exam should cover the same content as the basic math courses; the expediency of using SAT or ACT admissions test scores for placement is contrary to the purposes and design of these tests. Many basic math students overestimate the math they know or remember, and a student lacking proficiency should be strongly advised to take or assigned to a basic math course. A mechanism should be in place at registration to check that the student in fact enrolls. Of course, test scores like other predictors of achievement are imperfect, and any students who feel that their designated courses are inappropriate should be allowed to request a hearing or a retesting.

(3) The program should be responsive to the special needs of its students.

The curriculum should be organized into a course sequence sufficiently flexible so as to provide students in different situations with a reasonable chance to succeed as quickly as possible; an intensive arithmetic course should be available for the weakest student, a briefer arithmetic course for the marginally deficient who only needs a quick review, a separate course in elementary algebra for the student who knows only arithmetic, etc. Sections should be offered at times convenient to students, many of whom hold jobs and have family responsibilities. Courses must meet a sufficient number of contact hours a week -- at least four -- so as to allow practice in class. Where courses are self-paced, a calendar of minimum progress should be set to reduce the chances of procrastination. Even if the dominant mode of instruction is traditional lecture-recitation, individual student needs should be addressed outside the classroom by tutoring, CAI, etc. Instructors should be knowledgeable in alternative approaches to each topic, and attempt to gauge and to build on each student's previous mathematical experience. Many basic math students have poor study skills, so that the choice of textbook ought to take into account not only curricular fit but also such factors as quality of writing, readability level, and the extent to which skills are constantly being reinforced throughout by means of review exercises integrated with new material.

(4) <u>The program must be provided with sufficient resources to carry out its work.</u>

Unfortunately, many basic math programs do not have enough sections to accommodate all students designated as remedial. The college must allocate resources to provide the program with sufficient staffing and space. Enrollments in traditionally taught sections should be limited to 25 so as to allow for individual attention. The majority of sections should be taught by full-time math faculty who also teach non-remedial courses so as to maximize the continuity between remedial and non-remedial mathematics. The Math Lab must offer enough tutoring to meet student needs, and staffing should be funded in such a way as to provide continuity of employment to key personnel, not the practice on many campuses. Microcomputers, video setups and other educational technology should be made available in the Math Lab, for the use of those students who prefer these alternative modes of instruction or drillwork.

(5) The integrity of the program and of the college degree should be protected.

If the remedial requirement is to have teeth, each basic math course should have firm exit standards including a uniform final exam comparable to the corresponding placement subtest. This is especially important when one course leads to another in a sequence, or where part-time faculty, in whom one may not have the same degree of confidence as full-timers, predominate. Students should be required to demonstrate mastery of major topics (mastery learning), because of the cumulative and fundamental nature of the curriculum. At registration into client courses, including freshman mathematics, science, etc., students who have not fulfilled their remedial obligations should be screened out if these courses are to be taught at the college level; this may be a serious bone of contention where the jobs of physics faculty, for example, are at stake. Awarding degree credit for arithmetic or elementary algebra weakens the degree and should be avoided, although granting credit for such courses toward either full-time student status or for financial aid may be essential to enable the student to attend college.

Conclusions: Reducing the Need.

Having reviewed the rise of TYC basic math programs, the extent to which they are effective, and some suggestions for building successful programs, I would now like to turn to the question of whether the need for such programs will ever be reduced, and if so, under what circumstances this reduction can be brought about. The underlying question, of course, is how the level of mathematical preparation of students entering TYC's can be raised. Fortunately, there are a number of scenarios which might lead to that delightful prospect.

In the first of these scenarios, the one which will be opposed by those of us who believe in open-admissions for reasons of social equity and mobility, TYC's might adopt restrictive admissions criteria. Such a development is not unthinkable. Tight funding has already caused a national trend among many public state universities to toughen their entrance requirements and to get out of the remediation business.[13] Will this trend extend to the TYC? One possibility is that TYC's, for the very economic or political reasons that motivated the senior institutions, will move in the same direction. Another is that an even larger portion of weak students will be shunted to TYC's diverted from the increasingly exclusive four-year institutions.

In fact, some TYC's have in the past few years begun to move toward toughening their admissions standards.[14] Many others could conceivably choose this option in about a decade when there will be a turn-around in the demographic curve, and the number of eighteen-year-olds will begin to climb steadily [6]; at that time, it will be less risky for an institution to shrink or shift its target population.

In the second scenario for reducing the need for TYC math remediation, the high schools could begin to graduate students with stronger math skills. These days, the states seem to be tripping over one another in efforts to strengthen the exit standards of their high schools.[13] Such efforts reflect intense public and governmental dissatisfaction with the condition of the educational system, particularly the failure of the high school diploma to signify readiness either to enter the job market or to study at the post-secondary level. This commitment to strengthen the high school curriculum could lead, at least in some states or districts, to an across-the-board requirement that students in all tracks pass a course in elementary algebra. There are already efforts in this direction, efforts which all of us in the mathematics community need to support. For example, a National Academy of Sciences panel of business and education leaders recently noted that students who start work after high school need to master virtually the same basic academic skills as college-bound students, and recommended that all high-school students be required to take elementary algebra. [15] Likewise, the National Commission on Excellence has suggested that high schools require all students to take a minimum of three years of mathematics.[16]

Another approach to strengthening the de facto secondary math curriculum is through assessment programs. Many states and school districts already give their high school students competency tests which they must pass in order to earn a regular diploma. The mathematical content covered on these tests is generally restricted to elementary computational skills, often as applied to daily-life situations. In time, the competency tests already in place should improve the computational skills which students bring to the TYC, reducing the arithmetic component of basic math programs. Extending the test content to include a good deal of elementary algebra would be a boon for the TYC.

Persuasion might succeed where coercion fails. A selling job could coax high school students into taking more solid mathematics either to prepare for future employment [17], or to avoid future remedial obligations.[18]

Strategies might be developed which encourage high school students to learn and retain more of the mathematics which they study. Tapping the potential of microcomputers may one day allow students to explore concepts and to reinforce skills in ways not now dreamed of. The degree to which math skills are reinforced in other high school courses could be the focus of an arithmetic- or an algebra-across-the-curriculum movement. However, any attempt to raise the quality of math instruction in the high schools will involve finding ways to reverse the shortage of qualified, certified teachers, an effort to which the TYC's could conceivably contribute.

In the third scenario, students would be persuaded to matriculate in college right after graduating from high school, reducing the extent to which skills grow rusty. The state of the economy and the availability of financial aid could be key factors in discouraging students from taking a break between high school and college. However even if such efforts were wildly successful, the TYC's would still have to reckon with the needs of its older freshmen. At many TYC's, this population is already substantial in size; for example, a good fourth of the regular admittees at CUNY's TYC's enter at age 25 or above.[19]

In the final scenario for the contraction of TYC basic math programs, ignorance of traditional paper-and-pencil arithmetic could become less of a handicap for incoming students. Public confidence in the importance of conventional computational algorithms might, in time, erode to the point that weak students could simply be handed a calculator without suffering any stigma from their machine-dependence. TYC basic math programs would then focus on calculator-based skills, algebra, problem-solving, and a limited number of elementary geometric topics.

REFERENCES

[1] Fey, J. T., Albers, D. J., and Fleming, W. H. Undergraduate Mathematical Sciences in Universities, Four-Year Colleges and Two-Year Colleges, 1980-1981, Washington, D.C.: Conference Board of the Mathematical Sciences, 1981.

[2] Cross, P. Accent on Learning, San Francisco: Jossey-Bass Publishers, 1979.

[3] Lederman, M. J., Ryzewic, S. R. and Ribaudo, M. Assessment and Improvement of the Academic Skills of Entering Freshmen: A National Survey, Instructional Resource Center, CUNY, 1983.

[4] Clifford, R. "Devaluation, Diffusion and the College Connection: A Study of High School Transcripts, 1964-1981." National Commission on Excellence in Education (ED 228 244).

[5] Simon, K. and Grant, W. C. Digest of Educational Statistics, Washington, D. C.: Dept. of Health, Education and Welfare, 1969.

[6] Andersen, C. 1981-82 Fact Book for Academic Administrators, Washington, D. C.: American Council on Education, 1981.

[7] Akst, G. and Hecht, M. "Program Evaluation," in Teaching Basic Skills in College, San Francisco: Jossey-Bass Publishers, 1980.

[8] New Jersey Basic Skills Council Report on the Effectiveness of Remedial Programs in New Jersey Public Colleges and Universities, Fall 1982, 1983.

[9] Jorgensen, P. R. Survey of Remedial/Developmental Mathematics Programs at Two-Year Public Colleges, a doctoral dissertation presented in partial fulfillment of the degree requirements at Wayne State University, 1981.

[10] Office of the Vice President for Administration and Planning, Manhattan Community College Analysis of Academic Performance After Seven Calendar Semesters for First-time Freshmen and Advanced Standing Students Admitted in the Fall 1980 Semester, June 1984.

[11] Lavin, D., Murtha, J. and Kaufman, B. Long Term Graduation Rates of Students at CUNY, Office of Institutional Research and Analysis, CUNY, March 1984.

[12] Watkins, B. T. "Associate-Degree Programs 'Weak,' Community College Officials Say," Chronicle of Higher Education, September 1982, p. 10.

[13] Maeroff, G. I. "High Schools and Public Colleges Stiffen Requirements for Pupils," New York Times, February 7, 1983, p. 1.

[14] Watkins, B. T. "At Some Community Colleges, the 'Open Door' Begins to Close," Chronicle of Higher Education, September, 1982, p. 1.

[15] New York Times, "'Core' Skills Needed by Work-Bound Pupils," May 24, 1984.

[16] National Commission on Excellence, A Nation at Risk, Washington, D. C.: US Govt. Printing Office, 1983.

[17] Choike, J. R. Mathematics at Work in Society: Opening Career Doors, Washington, D. C.: American Statistical Association, 1980.

[18] Waits, B. and Leitzel, J. "Early University Placement Testing of High School Juniors," Mathematics in College, Fall 1984.

[19] Lavin, D. et al. Socioeconomic Origins and Educational Background of an Entering Class at CUNY, Office of Institutional Research, CUNY, 1983.

(DISCUSSION BEGINS ON P. 185.)

LIBERAL ARTS MATHEMATICS--

CORNERSTONE OR DINOSAUR ?

by

Karl Smith

SUMMARY

Liberal arts education in general, and liberal arts mathematics in particular, has a long history in American education, but since 1972 enrollments in liberal arts mathematics in the two-year colleges have dropped from 72,000 to a low of 19,000 in 1981. This paper explores some of the reasons for this decline and then looks toward the future and asks the question, "Can or should the old 'great ideas' mathematics course be revitalized to fit the needs of the students in the 1980s?" The thesis put forth in this paper is that it can be revitalized, but only if it takes on new form and includes a new orientation. The old liberal arts course will not only need to be modified to fit the career-oriented student, but also will need to integrate basic mathematics competencies including new technologies and be organized around the great ideas concept, rather than around specific basic skills. Liberal arts mathematics is an important course in the mathematics curriculum, and should continue to be taught. The liberal arts course should not have a fixed content, but should offer a variety of topics from which individual classes and instructors can choose. This choice should include statistics, computers, and discrete topics.

LIBERAL ARTS MATHEMATICS - CORNERSTONE OR DINOSAUR?

Karl J. Smith

History of Liberal Arts Mathematics

Liberal arts education can be defined as "that part of education which seeks to make people reflective and responsible; to relate art, music, and literature to their lives; to increase their understanding of the past, present, and future of the society of which they are members; and to bring them into the culture. Its roots are in the Greek ideal of liberal education, of educating people for participation in the polity." [1]

The concept of liberal arts education began in the medieval Greek universities with the study of grammar, rhetoric, logic, music, astronomy, geometry, and arithmetic, and included those ideas which were considered essential for the learned person. Later the curricula were expanded to include the classical languages, philosophy, and the natural sciences. By the nineteenth century the physical and social sciences were also included. During the nineteenth century, scholars began to associate themselves with specific academic disciplines and liberal arts took on a disciplinary form.

In 1924, Koos studied the curricula in fifty-eight

public and private junior colleges and found the liberal arts totaling three fourths of the curricula. Within the curricula, mathematics accounted for 7.4% of the total offer ings in those junior colleges [7]. In collegiate mathematics, liberal arts courses were developed to survey the underlying ideas of mathematics, and then to apply them to a wide variety of mathematical settings, such as algebra, geometry, probability, statistics, and logic. This emphasis in liberal arts courses continued through the 1960s and developed into what have been called the "great ideas" mathematics courses. These courses were built around the great ideas which revolutionized modern thinking, the great ideas which created major areas of study in the mathematical sciences, as well as many of the great ideas of famous mathematicians. Therefore, these liberal arts courses often had a definite historical flavor, and were aimed toward "developing an appreciation of mathematics," rather than toward teaching specific manipulative skills.

The Present State

In the early 1960s, most two-year colleges had a liberal arts orientation, serving as feeders to four-year colleges. In the 1960s we heard the students tell the schools "You have no authority to tell me what to study!" and the educators of the times agreed that their programs were not relevant. In the late 1960s, the word relevant meant applicable to living a more democratic, peaceful, communal, and ecological life. However, by the mid-1960s, there had been considerable change in the emphasis of the courses. New programs were introduced in the occupational/technical areas — in data processing, dental hygiene, electronics, practical nursing, automotive mechanics, accounting, bricklaying, carpentry, and police and fire science.

But during the 1970s it became obvious that many persons with liberal arts backgrounds were unable to parlay their education into jobs, or to incorporate their education into the jobs at which they were employed. There was a cry for meeting student needs and providing skills that would lead to employment. The consumerism of education in the 1970s became skill oriented and between 1975 and 1980 there was a 74% drop in liberal arts mathematics enrollments (see Table 1).

TABLE 1 ENROLLMENTS IN LIBERAL ARTS MATHEMATATICS IN TWO-YEAR COLLEGES IN THE UNITED STATES [5]

Year	Enrollment
1966-67	22,000
1970-71	57,000
1975-76	72,000
1980-81	19,000

Among the reasons for the decline in liberal arts mathematics courses was the desire on the part of the student to obtain a "real job" and the corresponding specific required mathematics courses from algebra through calculus, the increased demand of specialization, the decline of the number of persons entering elementary education, and the mood of the 1960s. By the early 1970s the catchword became <u>individualism</u> and we were no longer preparing people to become educated members of a free society, but instead were preparing them solely for a job. In an article on liberal education, C. J. Hurn summarized the feeling of the time:

"Lacking any consensus as to the content of liberal education, and lacking confidence in their prescriptive authority -- as the catchphrase puts it, 'to impose their values upon others' -- educators were in a weak position to mount a defense of anything other than an educational supermarket, where customer preferences, in the middle and late 1970s at least, were clearly for the more immediately utilitarian and basic items on the shelf." [6]

Currently, full-time-equivalent enrollments in occupational/technical programs lead enrollments in college transfer programs. This is a reversal of the trend prior to 1970 as shown in Figure 1.

Figure 1 Percentage of Full-Time-Equivalent Enrollments

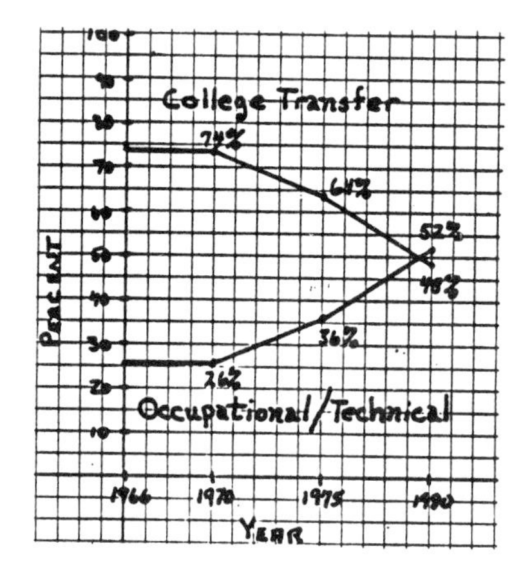

Source: <u>Projections of Education Statistics to 1986-87</u> and The Conference Board of Mathematical Sciences questionnaire data for 1980 (Washington: The Conference Board of Mathematical Sciences, Report of The Survey Committee, Vol. VI, 1981), p. 83.

As you can see from Figure 1, less than half of two-year college students are now enrolled in transfer programs. This reversal in the type of students being served by two-year colleges has led to a reevaluation of the traditional liberal arts program in general, and to a reassessment of liberal arts mathematics in particular.

Look Toward the Future

Enrollments in mathematics, in general, are increasing at the same rate as college enrollments (see Figure 2), and

Figure 2 Enrollments in Two-Year Colleges

Year	Total Students	Mathematics
1966	1,464,099	348,000
1970	2,499,837	584,000
1975	4,069,279	874,000
1980	4,825,931	1,048,000

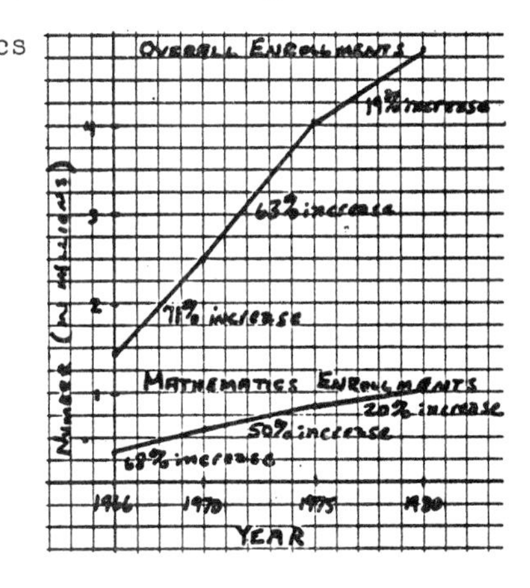

Source: <u>1981 Community, Junior, and Technical College Directory</u>, American Association of Community and Junior Colleges, One Dupont Circle, N.W., Washington, D.C. 20036 and The Conference Board of Mathematics Sciences, Report of the Survey Committee, 1981, op. cit., p. 74.

therefore we can conclude that students who, in the past, enrolled in liberal arts courses are now enrolling in other mathematics courses. Given the decline (and some are saying demise) of liberal arts mathematics in the United States, what is its future? Will students continue to avoid liberal arts mathematics and enroll in other courses? Should the course be scrapped and replaced by the new wave of courses, such as discrete mathematics, finite mathematics, computer programming, or even by the "main line" mathematics courses of algebra, trigonometry, and calculus? Can the old "great ideas" mathematics course be revitalized (or resurrected) to fit the needs of the students in the 1980s? A great ideas course is a course that is built around the great ideas of mathematics rather than around specific skills. For example, a great ideas course would focus on the nature of geometry, and what kinds of geometry are possible, rather than the specifics of areas, volumns, and angles.

A reorientation of the liberal arts curriculum is definitely necessary. But at the same time we cannot lose sight of what this course has always been -- a fun course. Fun for the students and also for the instructor. This is one course with which we have the opportunity to have fun and to present those great ideas which have had such a profound impact on our civilization. William Lucas, in his paper Problem Solving and Modeling In the First Two Years states [8] : "Two of the major reasons why mathematics, minus computer science, has rapidly been losing student interest is that we have stripped the 'fun' out of doing mathematics and we have furthermore neglected to convince students in a meaningful way that mathematics is a contemporary, dynamic and currently relevant subject." He then goes on to build a case for a discrete mathematics course, but I think you could also build a case for a liberal arts course as well. The stength of liberal arts has been, and should continue to be, a course which offers a variety of topics from which individual classes and instructors can choose topics suitable for that particular class. For example, such a course should include, but not be limited to, algebra, geometry, statistics, computers, and discrete topics. It should serve not only as a stimulus and vehicle for those adults who are returning to school looking for more than an interlude between high school and a profession, but it should also serve those who are looking for basic skill-building mathematics courses. At a time when students are saying, "Don't bother me with something that is not part of the basic required curriculum," can we tell them, "Yes, manipulations and equation-solving skills <u>are</u> important, but mathematics is much more than manipulations and problem solving"? To do away with a historically oriented great ideas mathematics course would allow the pendulum to swing way too far in the other direction. We need, instead, a new type of liberal arts course, one that not only focuses on the great ideas of mathematics, but also integrates those great ideas into a course satisfying a list of basic mathematical competencies as well. If the liberal arts course is to survive, it will have to be modified to fit the career-oriented student. It will need to be modified to satisfy the competency requirements now being imposed by several states.

Consider, for example, the College Level Academic Skills Test (CLAST) in mathematics which is being required for graduation in the state of Florida. These requirements of <u>basic skills</u> in arithmetic, geometry, algebra, statistics, logic, and computer technology are applied to specific skills in the following categories:

I. Algorithmic Processes
II. Concepts
III. Generalizations
IV. Problem Solving

These basic competencies do not seem to be built around the old "great ideas in mathematics" of the traditional liberal arts course. But what course in the traditional mathematics curriculum is best able to meet these needs? In California, two-year colleges must soon certify that their graduates have demonstrated a basic competency in mathematics. Are these requirements of basic competency to be interpreted as separate courses such as algebra or geometry, or should there be an integrated course developed to meet these requirements? Perhaps the liberal arts course of the 1960s and 1970s can satisfy the needs of the future by assuming a dual role. This dual role is discussed by Cohen and Brawer [2]:

> For the sake of their students and communities, the community colleges should maintain a place in higher education, but a reorientation is required. One area of possible integration of the liberal arts with career education is a merger of principles stemming from both the humanities and the sciences. Technology is ubiquitous; students would have little difficulty understanding generally how the history, politics, ethics, sociology, and philosophy of science and technology affect their world. Those who would be more than mechanics would attend to the fundamental assumptions undergirding what scientists and technologists do.

In attempting to fit the liberal arts great ideas course into the dual role of maintaining basic skills within an environment that provides for reading, reflecting, and discussing, the colleges should not neglect the contrasting modes of teaching the liberal arts and the occupational courses. The traditional liberal arts mathematics course is taught by an instructor in a room with chairs, a chalkboard, and perhaps an overhead projector. If the class size permits, the instructor interacts with the students to stimulate the thought process and set up a dialogue of learning. On the other hand, the occupational courses have required labs, shops, and equipment. Have we simply accepted the idea that career teaching costs more because it has evolved from a history of apprenticeships in work settings? Instructors of these occupational courses maintain that their students must practice the craft, and not merely talk about it.

For many years we have asserted that "mathematics is not a spectator sport," but what does that mean for schools and instructors of the future? Should we continue to teach liberal arts mathematics at a chalk board, or should we move out into the real world to give our students a variety of hands-on experiences in the lab and in the community? For example, computer assisted instruction is now common in many parts of the country. Community projects involving business and industry can be designed for or by students of these new liberal arts courses in order to help the students gain firsthand experience in working with business and industry. How does mathematics relate, for example, to banks and banking? Can students taking this type of course be used to assist in the gathering of demographic or other information useful to city or state governments? How can computers be integrated into everyday activities? Can we demonstrate the great ideas of mathematics in a problem-solving, skills-oriented setting? New available technologies must be incorporated into the new liberal arts course. Can we change the modes of teaching liberal arts mathematics in the future as it assumes the new function of meeting basic competencies, perpetuating the great ideas of mathematics as well as integrating new available techonologies? Do we, the mathematical community, let the great ideas mathematics course of the past become the dinosaur of the 1990s, or do we change, modify, and build it into a basic skills course which <u>does</u> meet the competency requirements of the 1980s? There <u>is a need</u> for building basic skills in a survey-type course covering arithemtic, algebra, and geometry, as well as topics from logic, probability, statistics, and computer technology. We <u>can</u> use such a course as the cornerstone for meeting the job-oriented demands of today's, and tomorrow's, students. The appendix gives an outline for such a course.

Recommendations

1. Liberal arts mathematics is an important course and should integrate basic mathematics competencies including new technologies and be organized around the great ideas concept, rather than around specific basic skills.
2. Liberal arts mathematics should not have fixed content, but should offer a variety of topics from which individual classes and instructors can choose. However, this choice should include statistics, computers, and discrete topics.

A Sample Course Outline

How can a single course incorporate basic mathematical competencies with the great ideas of the old liberal arts course presented with a historical perspective? The following list of topics is only one possibility and is offered for further discussion.

I. Logic
   A. Sets
      1. Unions, intersections, complements
      2. Combined operations
      3. Venn Diagrams
   B. Inductive Reasoning and Patterns
      1. Arithmetic sequences
      2. Geometric sequences
      3. Series
   C. Deductive Reasoning
      1. Fundamental connectives: and, or, not
      2. Additional connectives and logical translation
      3. Proof

II. Arithmetic
   A. Historical Numeration Systems
      1. Repetitive system (such as the Egyptian)
      2. Positional system (such as the Babylonian)
      3. Hindu-Arabic system

- B. Natural Numbers
  - 1. Properties
  - 2. Exponents
- C. Integers
- D. Rational Numbers
  - 1. Fractions
  - 2. Decimals
  - 3. Percents
- E. Ratio and Proportion
- F. Irrational Numbers
  - 1. Square roots
  - 2. Pythagorean theorem
- G. Real Numbers

III. Algebra
- A. Mathematical Systems
- B. Algebraic Simplification
- C. Linear Equations
- D. Linear Inequalities
- E. Systems

IV. Geometry
- A. Geometrical Figures
- B. Systems of Measurement
  - 1. U. S. system
  - 2. Metric system
- C. Measuring Length and Area
- D. Angles
- E. Euclidean Geometry
  - 1. Proofs
  - 2. Parallel lines
- F. Volume and Surface Area
- G. Coordinate Geometry
- H. Lines and Half-planes

V. Probability and Statistics
- A. Combinatorics
  - 1. Counting
  - 2. Permutations
  - 3. Combinations
- B. Probability Models
- C. Conditional Probability
- D. Mathematical Expectation
- E. Frequency Distributions
- F. Descriptive Statistics
  - 1. Mean, median, mode
  - 2. Variance, standard deviation
- G. The Normal Curve
  - 1. Standard normal
  - 2. z-scores
- H. Sampling

VI. Computer Technology
- A. History of Computers
- B. Calculators
- C. Uses and Misuses of Computers
- D. BASIC Programming
- E. Loops and Repetitive Processes

REFERENCES

[1] Cohen, Authur M. and Brawer, Florence B., The American Community College (San Francisco: Jossey-Bass Publishers, 1982), p. 283.

[2] Ibid., p. 304

[3] Conference Board of Mathematics Sciences (CBMS), Report of the Survey Committee 1981 Community, Junior, and Technical College Directory, American Association of Community and Junior Colleges, One Dupont Circle, N.W., Washington, D.C. 20036 and p. 74. (Washington: The Conference Board of Mathematical Sciences, Report of The Survey Committee, Vol. VI, 1981), p. 83.

[4] Conference Board of Mathematics Sciences (CBMS), Report of the Survey Committee Projections of Education Statistics to 1986-87 and Conference Board of Mathematical Sciences questionnaire data for 1980. (Washington: The Conference Board of Mathematical Sciences, Report of The Survey Committee, Vol. VI, 1981), p. 83.

[5] Fey, James T., Albers, Donald J., and Fleming, Wendell H., Undergraduate Mathematical Science in Universities, Four-Year Colleges, and Two-Year Colleges, 1980-81, Volume VI (Washington: Conference Board of The Mathematical Sciences, Report of the Survey Committee, Vol. VI, 1981), p. 83.

[6] Hurn, C. J., "The Prospects for Liberal Education: A Sociological Perspective," Phi Delta Kappan, 1979, Vol. 60, No. 9, p. 632.

[7] Koos, E. L., The Junior College, (Minneapolis: University of Minnesota Press, 1924), p. 29.

[8] Ralston, Anthony and Young, Gail S., The Future of College Mathematics, (New York: Springer-Verlag, 1983), p. 43.

(DISCUSSION BEGINS ON P. 185.)

STATISTICS IN THE TWO-YEAR COLLEGE CURRICULUM

by

Ann Watkins

SUMMARY

In spite of the recommendations of major reports on curriculum reform, calculus still dominates the mathematics curriculum. As a result, statistics is taught to a relatively small percentage of students. Statistics is not required for high school graduation, for college admission, or for success on standardized tests, so very few students have coursework in probability and statistics in high school. Those students who then take a college-level statistics course do not have the appropriate prerequisite skills. The mathematics courses they have taken have prepared them for calculus, not for statistics.

Two-year college mathematics teachers have not been trained to teach statistics and as a result over half do not feel well-qualified to do so. Mathematics teachers are largely unaware of the changes in the statistics curriculum that statisticians have been recommending. These changes include using real data, teaching data analysis techniques, emphasizing understanding of statistical concepts, and spending less time on computational formulas.

If an understanding of statistics is to become part of the minimal knowledge required of two-year and four-year college graduates, the mathematics curriculum will have to be designed so that it emphasizes statistics as well as calculus and mathematics departments will have to demand as much training of teachers of statistics as is required of calculus teachers.

STATISTICS IN THE TWO-YEAR COLLEGE CURRICULUM

Ann E. Watkins

I have been asked to "make a case for statistics" in the two-year college. No one needs to "make a case" for computers or for calculus or for trigonometry, so why for statistics? It is not as if the case has not been made before. Nearly every major report on curriculum reform of the last twenty-five years states that a knowledge of statistics is indispensable in daily life and recommends that statistics be taught throughout the grades. For example, the Committee on the Undergraduate Program in Mathematics (CUPM) report on minimal mathematical competencies for college graduates says that four-year colleges and universities "should expect graduates to understand and be able to use some elementary statistical ideas,.... This also applies to two-year college students in university parallel curricula." (CUPM Panel, 1982). Most recently, the National Science Board Commission on Precollege Education in Mathematics, Science, and Technology stated that "elementary statistics and probability should now be considered fundamental for all high school students." (National Science Board, 1983).

Mathematicians agree that statistics is one of the most useful topics that we can teach. In CUPM's survey of (mostly) mathematicians, respondents chose the mathematical topics from a forty-item list that they thought should be required of all college graduates. Elementary statistics was chosen by 56%. Only basic arithmetic skills, area and volume of common figures, linear equations, and algebraic manipulations were chosen more often. When asked what standard courses should be required, probability and statistics was mentioned second most often, following college algebra.

Many eloquent statements have been made for requiring statistics at every level of schooling. There is no need to repeat the arguments here. They are based on the fact that it is important for a citizen to be able to understand articles like the following:

> In the latest Gallup survey, 54% approve of Reagan's handling of his presidential duties while 37% disapprove.... The last time Reagan had a significantly higher job performance rating was in August, 1981, when 60% approved and 29% disapproved.
>
> Reagan's strongest showing (although a statistical tie) in this series was recorded last December, when he received 51% of the vote to Mondale's 44%.
>
> The latest results are based on interviews Jan. 13 to 15 with 1,139 adults.... The error attributable to sampling and other random effects could be 4 percentage points in either direction. (Los Angeles Times, 1/29/84)

Unfortunately, a large part of our elementary algebra - geometry - intermediate algebra - trigonometry - precalculus sequence is appropriate only for the student who will make it all of the way through to calculus. A great deal of time is spent on calculus-related skills such as factoring quadratics and manipulating rational expressions. Since the vast majority of our two-year college students never transfer to four-year colleges, much of the mathematics we now teach will be of no use to them. If the goal of our mathematics sequence was statistics instead of calculus, then those students who drop out along the way would still have learned something useful.

How Much Statistics is Being Taught in Two-Year Colleges?

With such strong and virtually unanimous support for statistics, why must I make yet another case for it? Because, in spite of all that has been advocated, statistics is still not being taught to a large number of students.

In two-year colleges, we offer a one-semester course for those students who are required to take it as part of their major. It is rare that statistics is a required part of general education for transfer students, as advocated by CUPM, although it is often an unused option.

Statistics is a small part of total mathematics enrollments. The 1980-1981 Conference Board of the Mathematical Sciences survey found that only 3% of all mathematics enrollments in two-year colleges were in statistics - virtually the same as in 1970. This is about one-third of the enrollment of calculus! The percentage of two-year colleges offering a course in statistics actually declined from 41% in 1970 to 28% in 1980. The percentage offering a course in probability with statistics declined from 16% to 14%. Enrollments in universities and four-year colleges are similar. Out of a total of 1,966,000 enrollments in undergraduate mathematical science courses in 1980, 104,000 were in elementary statistics and 45,000 were in intermediate and advanced statistics. In comparison, there were 518,000 students in analytic geometry/calculus (Fey el al., 1981).

Since mathematics departments are teaching so few sections of statistics, are students learning statistics from other departments? There has been a large increase in the enrollments in mathematics courses taught outside mathematics departments until they now equal 13% of the total math enrollment in two-year colleges. However, only about 4.5% of outside enrollments are in statistics and probability. These courses are taught by business (42%), occupational programs (33%), natural sciences (17%), and social sciences (8%) (Fey et al., 1981). The 1980 Center for the Study of Community Colleges survey found that probability and statistics courses made up 5% of total mathematics and computer science sections (inside and outside of mathematics departments) listed in schedules of classes (Beckwith, 1980). These figures are comparable to the ones from the American Mathematical Society survey of four-year colleges offering at most a masters degree (Rung, 1984).

In short, we do not have a good estimate of the percentage of two-year college students who have taken a course in statistics, but the number is small compared to calculus enrollments.

Why Isn't Statistics Being Taught?

Why is so little statistics taught within mathematics departments to so few students? One reason may be that statistics is more than just mathematics. Except for the simple data cooked up for textbook examples and exercises, statistical problems seldom have one correct answer or one correct way of proceeding. Good statistics teaching requires a knowledge of the "real world" in which experiments must be designed and surveys carried out in the absence of a perfectly random sample and in the presence of "bad" data. Mathematics teachers are not always comfortable with these difficulties. The typical two-year college teacher has not been trained to teach statistics. If a two-year college teacher has taken a course in statistics at all, it is most likely an upper division course in mathematical statistics. It is rare to find a two-year college faculty member with a degree in statistics (Fey et al., 1981). A 1977 survey of about 10% of the two-year college mathematics teachers in the U.S. found that 74% of those holding a doctorate felt "entirely secure about my qualifications to teach" statistics, but only 46% of those with a masters plus one year and 42% of those with a masters felt this confident (McKelvey, et al., 1979). In contrast, the percentages who felt entirely secure about their qualifications to teach calculus were 91%, 86%, and 85%, respectively.

A second reason that there is not much statistics being taught is that students are unprepared to take it. If a student has had no prior experience with statistics, the introductory college-level statistics course is very difficult. It contains a large amount of material and tricky logical arguments. Probability is the most notorious topic in this regard. In the two or three weeks that can be allotted to it, students with no prior training are not able to acquire any indepth knowledge of the subject and usually resort to memorizing definitions.

How Much Statistics Do Students Learn in High School?

How much statistics and probability can we expect a student to know when he or she enters college? Very little, it appears. Unfortunately, the situation has not changed since NACOME reported in 1975,

> While probability instruction seems to have made some progress, statistics instruction has yet to get off the ground ... At the high school level probability topics in Algebra I and II texts are commonly omitted. A one semester senior course in probability and statistics has gained only a small audience of the very best students. Furthermore, this course places a heavy emphasis on probability theory, with statistics,

if treated at all, viewed as merely an application of that theory. Though National Assessment gives reasonable attention to probability and statistics objectives, current commercial standardized tests do virtually nothing with these topics.

How many students have taken a one-semester course in statistics and probability in high school? The latest Science Education Databook from NSF says that only 2.7% of 17 year-olds report having taken a course in statistics and probability (Directorate for Science Education, 1980). Of the 42 high schools in New Hampshire, only five offer a course in statistics (Prevost, 1983).

The problems confronting the high school teacher who wants to offer a statistics course are considerable. Unbelievably, there is no up-to-date statistics textbook for secondary school students on the market. Consequently, most high school teachers use a book written for college, typically Newmark (1977). Fortunately, this situation may be changing. The American Statistical Association - National Council of Teachers of Mathematics Joint Committee on the Curriculum in Statistics and Probability has recently received a grant from NSF to write, field-test, and distribute four booklets for secondary school students on exploring data, probability, simulation, and statistical inference.

The second problem confronting the high school teacher is that there is no popular support for requiring such a course (NCTM, 1981). The number of students themselves electing such a course remains small as statistics is not required for high school graduation, for college admission, or for success on standardized tests. For example, a typical SAT will contain at most one probability question and two or three questions concerning the average. The College Board does not want to disadvantage students who have not been exposed to formal study of probability and statistics (Statistics Teacher Network Newsletter, September 1983).

If high school students are generally not taking a semester course, how much probability and statistics has been included in other courses? A survey of approximately 350 high schools in Wisconsin provides the answer. The percentage of schools which allot more than three weeks in the total high school program to statistics declined from 26% in 1975 to 23% in 1983. In 1975, 43% allotted more than three weeks to probability. This declined to 34% in 1983 (Williamson, 1983).

Since such a low priority is placed on statistics and probability in the high school curriculum, it is not surprising that the National Assessment of Educational Progress has found that students do not know very much about statistics. For example, the terms "mean," "median," and "mode" are unfamiliar to a majority of 17 year-olds (Carpenter, et al., 1980). Less than 30% of 17 year-olds could answer the following question.

> In three tosses of a fair coin, heads turned up twice and tails turned up once. What is the probability that heads will turn up on the fourth toss? (Hope and Kelly, 1983)

As students do not learn much about statistics in high school, two-year colleges do not require any knowledge of probability or statistics for admission to a college-level statistics course. According to the Center for the Study of Community Colleges data, 74% of the courses listed a prerequisite: elementary algebra by 16%, placement test by 3%, intermediate or college algebra by 40%, analytic geometry or trigonometry by 9%, business or technical math by 7%, consent of instructor by 4%, and another course in the discipline offering the statistics by about 10% (Beckwith, 1980). Clearly there is no nationwide agreement on the necessary prerequisites, except that no prior study of the subject is required.

What Do We Teach in Elementary Statistics?

A survey of the most popular elementary statistics books reveals a tightly organized, highly sequential progression through descriptive statistics, probability, probability distributions, estimation, and hypothesis testing, with other largely optional topics tacked on at the end. The course is so beautifully constructed and mathematically logical that no one wants to change it. Although the topics are generally the right ones, the emphasis is on computation. Two-year colleges do not typically have MINITAB, SPSS, or other statistical packages that students can use to perform routine computations. Consequently, a student's success is judged by his or her ability to compute various statistics and to follow algorithms to see if the statistics are "significant." Everything is very structured. One enterprising teacher even made up a large flow chart for her students so that when they read a word problem, they will be able to select the correct statistical test. The exercise sets follow the model of elementary algebra texts. They give an example and then provide a bunch of exercises just like the example. Very few exercises test for any understanding. But then, how much can we ask of students who have never seen the subject before?

What Should We Teach?

Statisticians have been telling mathematics teachers for some time that we are presenting an out-of-date approach to statistics. They say they do not very often use the techniques we teach. Several introductory statistics books have recently been written by statisticians in the spirit of the "New Statistics." These include Ehrenberg (1982); Freedman, Pisani, and Purves (1978); Haack (1979); Koopmans (1981); Lindgren and Berry (1981); Moore (1979); Mosteller, Fienberg, and Rourke (1983); and Nemenyi et al. (1977).

How are these texts different from the ones we have been using? First, they emphasize understanding of statistical concepts and deemphasize computation. For example, here is an exercise from Freedman, Pisani, and Purves.

> An investigator develops a regression equation for estimating the weight of a car (in pounds) from its length (inches). The slope is closest to
>
> | | | |
> |---|---|---|
> | 3 pounds per inch | 30 pounds per inch | 300 pounds per in. |
> | 3 inches per pound | 30 pounds per sq.in. | 300 cm. per kg. |

Here is another from Moore.

> A psychologist speaking to a meeting of the American Association of University Professors recently said, "The evidence suggests that there is nearly correlation zero between teaching ability of a faculty member and his or her research productivity." The student newspaper reported this as "Professor McDaniel said that good teachers tend to be poor researchers and good researchers tend to be poor teachers."
>
> Explain what (if anything) is wrong with the newspaper's report. If the report is not accurate write your own plain-language account of what the speaker meant.

The type of thinking a student must do in order to answer these questions is quite different from the usual exercises which ask him or her to compute the slope of the regression line or the correlation coefficient.

The second major difference between the New and Old Statistics textbooks is that the New Statistics textbooks use real data almost exclusively. As Mosteller, Fienberg, and Rourke state in their preface, "the tang of reality adds zest to the learning process." They provide real data about questions such as, "Do judges in the Olympics give higher scores to athletes from their own countries?" and "How long a chain of acquaintances is needed to connect one person in the U.S. to another?". Nemenyi et al. give the original data from Michelson's 100 measurements of the speed of light. The use of real data in class not only is more fun, but it also prepares students for the messy data they are sure to encounter later.

Finally, the New Statistics textbooks include data analysis. Data analysis is a set of descriptive techniques combined with a flexible attitude. Students are encouraged to use simple and highly visual techniques of displaying data in order to generate hypotheses, in contrast to the usual approach of testing preconceived hypotheses.

Mathematical Statistics in the First Two Years?

This paper has not discussed statistics courses with a calculus pre-requisite as these are not currently taught in two-year colleges. However, such a course has been proposed for lower division and so two-year college teachers should be preparing to teach it. For further information, see CUPM (1981) and Ralston and Young (1983).

Conclusions

In spite of the recommendations of major reports on curriculum reform, calculus still dominates the mathematics curriculum. As a result, statistics is taught to a relatively small percentage of students. If statistical literacy is to become a fundamental goal of schooling, the mathematics curriculum will have to be redesigned so that it emphasizes statistics as well as calculus.

REFERENCES

Beckwith, Miriam M. Science Education in Two-Year Colleges: Mathematics. Los Angeles: Center for the Study of Community Colleges and ERIC Clearinghouse for Junior Colleges, UCLA, May 1980 (ERIC #187 386).

Carpenter, Thomas P. et al. Results of the second NAEP mathematics assessment: Secondary school. Mathematics Teacher, 1980, 73, 329-338.

Committee on the Undergraduate Program in Mathematics (CUPM). Recommendations for a General Mathematical Sciences Program. Mathematical Association of America, 1981.

CUPM Panel. Minimal mathematical competencies for college graduates. American Mathematical Monthly, 1982, 89, 266-272.

Conference Board of the Mathematical Sciences National Advisory Committee on Mathematical Education. Overview and Analysis of School Mathematics Grades K-12. Washington, D.C.: CBMS, 1975.

Directorate for Science Education. Science Education Data Book. Washington, D.C.: National Science Foundation, 1980.

Ehrenberg, A. S. C. A Primer in Data Reduction. New York: John Wiley & Sons, 1982.

Fey, James T. Mathematics Teaching Today: Perspectives from Three National Surveys. Reston, VA: National Council of Teachers of Mathematics, 1981.

Fey, James T., Albers, Donald J., & Fleming, Wendell H. Undergraduate Mathematical Sciences in Universities, Four-Year Colleges, and Two-Year Colleges, 1980-1981. Volume VI of the report of the CBMS survey committee. Washington, D.C.: CBMS, 1981.

Freedman, David, Pisani, Robert, & Purves, Roger. Statistics. New York: W. W. Norton, 1978.

Haack, Dennis G. Statistical Literacy: A Guide to Interpretation. Boston: Duxbury, 1979.

Hope, Jack A. & Kelly, Ivan W. Common difficulties with probabilistic reasoning. Mathematics Teacher, 1983, 76, 565-570.

Koopmans, Lambert H. An Introduction to Contemporary Statistics. Boston: Duxbury, 1981.

Lindgren, Bernard W. & Berry, Donald. Elementary Statistics. New York: Macmillan, 1981.

McKelvey, Robert et al. An inquiry into the graduate training needs of two-year college teachers of mathematics. Missoula MT: Rocky Mountain Mathematics Consortium, 1979.

Moore, David S. Statistics: Concepts and Controversies. San Francisco: Freeman, 1979.

Mosteller, Frederick, Fienberg, Stephen E., & Rourke, Robert E. K. Beginning Statistics with Data Analysis. Reading, MA: Addison-Wesley, 1983.

National Council of Teachers of Mathematics. Priorities in

School Mathematics: Executive Summary of the PRISM Project. Reston, VA: NCTM, 1981.

National Science Board Commission on Precollege Education in Mathematics, Science and Technology. Educating Americans for the 21st Century. Washington, D.C.: National Science Foundation, 1983.

Nemenyi, Peter et al. Statistics from Scratch. Pilot ed. San Francisco: Holden-Day, 1977.

Newmark, Joseph. Statistics and Probability in Modern Life. New York: Holt, Rinehart and Winston, 1977.

Prevost, Fernand J. Statistics and probability in New Hampshire schools. Statistics Teacher Network Newsletter, December 1983.

Ralston, Anthony and Young, Gail S. The Future of College Mathematics: Proceedings of a Conference/Workshop on the First Two Years of College Mathematics. New York: Springer-Verlag, 1983.

Rung, Donald C. Employment of mathematical sciences doctorates, faculty mobility, nonacademic employment and enrollments, fall 1983. Notices of the American Mathematical Society, 1984, 31, 146-151.

Williamson, Bruce. Probability and statistics in Wisconsin. Statistics Teacher Network Newsletter, September 1983.

(DISCUSSION BEGINS ON P. 185.)

DISCUSSION

ENDANGERED CURRICULUM ELEMENTS

1. Remediation: A Problem in Allocation of Resources

2. Liberal Arts Mathematics--Dying?

3. Statistics--For Whom?

REMEDIATION: A PROBLEM IN ALLOCATION OF RESOURCES

Leitzel: Near the end of his paper Geoffrey Akst talks about how we're going to put ourselves out of the remediation business which I think is another reason why remediation should be kept in the mathematics department. When it goes outside the math department, you have remediation kingdoms develop for the sake of preserving this kind of instruction, and it becomes good to have remedial students. When it's in the math department, that temptation is smaller.

Gordon: I have always had the feeling that remediation should be viewed as a short-term commitment --the kind of thing that you do over the next ten years or so while you are seeking other routes to eliminate it in the long term. The problem is that most mathematicians tend by their very nature to be passive individuals. We sit back and we wait for the students to show up on our doorstep. Occasionally, we go out with butterfly nets to grab whichever ones are loose, but we two-year college faculty don't take an active role in preparing them for college _before_ they get to college. I think that one of the big challenges that should face us over the next decade is to greatly strengthen ties with the secondary schools.

Page: Education is a socio-political process. The City University of New York has a tremendous number of minorities, disenfranchised students, and older people. My feeling is that if we didn't accommodate people in classrooms and spend the money for them that way, then that money would be spent on social programs that address violence, vandalism, crime, and any other number of things that would be concomitant with a lack of educational opportunities. There's a tremendous economic toll for this allocation of resources, and it's skewed in a certain direction. On the other hand, I don't know what the answer is. I agree with Geoffrey Akst that two-year colleges should not give credit for intermediate algebra or plane geometry, because after all that's only an incentive not to take it in high school. Why should you take it in high school? But by not remediating, what happens? You decrease that window of opportunity, that revolving door just spins around a little faster to knock the student out. The education is not as user-friendly, and therefore, the point is that you are losing more people. Ideally, we would like to buy opportunity for everybody, but the question then becomes: at what cost, at what allocation of resources, to whom, and depriving whom, and what happens if we don't do it?

Rodi: And what allocation of resources to whom? That's the question I always ask. Given enough time, enough money, and enough individual attention, you can take almost anybody and bring him or her to a certain level of education. But in doing that, you ask yourself, "How long has it taken that person to get there? Do they have enough productive years left to justify the investment?" And more importantly, "Could those same hours and that same money be allocated to a larger group of people to get more people ready to do more at an earlier stage?"

Liberal Arts Mathematics--Dying?

Cohen: Why would you raise the level of this course? You say the enrollments have declined from 72,000 to 19,000; it sounds like it is about to fall off the edge anyway.

Smith: We really need to make it a different course from what it was. I think the mood of the times, as I see it, is one of increased demands on our students, one of transferability for one thing. I think the old liberal-arts course is not being accepted by four-year schools in a variety of different places. Across the country, I see states saying that it, to be a transferable course, must be at the college algebra level, or at least have a prerequisite of intermediate algebra.

Renz: It seems to me by your analysis and also by that of others, that one of the reasons that this course succeeded is that it lacked a fixed position in the curriculum; that is, it is what is called a "terminal course." You didn't have to worry about making sure that you covered all the key items. Now, by injecting exactly that element, namely, saying that we are going to force this course to cover certain specific items, you are removing the very freedom which made the course a pleasure both for the teachers and for the students.

I think this course is surfacing under other names; it is a course directed at overcoming math avoidance and math anxiety, and many other sorts of things. But I think that the proposed remedy that you are putting forth is simply antithetical to the very nature of the essential qualities of the course.

Case: The Math Appreciation Panel of CUPM did an extensive study of these courses. Jerry Goldstein chaired that Panel. The report is two or three years old by now, and it first makes the point that if it (liberal arts math) is a listing of particular skills, it is NOT a math appreciation course. Secondly, the college-level academic skills test of the state of Florida lists fifty-six skills. There was never any intent that those fifty-six skills be taught in the colleges. These are things that a majority of faculty members from all disciplines in the state of Florida at universities and two-year colleges voted were necessary for success in their disciplines. Some skills won in the balloting, when skills essential to them lost. So you see, we cannot teach one without teaching the preceding one which lost in the balloting.

Smith: But I think the point that Peter Renz made is the point that I'm making. I don't feel that you list a set of skills or competencies, regardless of the intent of the people making up the list. I don't think that defines a "great ideas" course. I do not feel that a list of competencies defines or makes up such a course.

Renz: Your point really is that you are talking about a different kind of course which might reach the same students.

Kaput: I worry about thinking again of basic skills in the sense of algebra manipulation types of skills and so on. That's just the wrong way to be thinking about basic skills for people who are going to spend their lives in the twenty-first century. We have to broaden our conception of what constitutes mathematical thinking and what constitutes mathematics. We should be thinking about general sorts of thinking skills that include such things as the management of complexity, how to find patterns in data, how to manipulate data using a variety of representations, how to know when one representation of data is to be preferred over another, and how to use that ability. Generating and testing hypotheses--that ability--I think is critical, particularly since you are going to have computers at your fingertips in the future that will allow you to generate and test hypotheses. You've got to be able to think in those styles. Those are the critical basic skills and they, in fact, are preceded by some other basic skills that Larry Curnutt mentioned earlier--organizing data, and pulling data together. That's the sort of thing that should be done at a very early age because by the time students are in junior high school they are already going to be manipulating data in a fairly sophisticated way, so that when they are at the secondary level they're going to be playing around with the sorts of things that Joan Leitzel is talking about and a lot more, to be sure. Instead of just using calculators as the instrument for generating and testing hypotheses, you are going to have much more computer power. You cannot regard the existing curriculum as sacrosanct; I think it's a serious mistake if you buy into the remedial courses and set up elaborate systems to deal with them. In the long run, you're co-opting the real solution.

Rodi: I see this course (liberal arts math) as dying because nobody is going to be made to take it; and when they have a choice, it's not clear that it will be the best choice even for the student. Suppose that a student who is going to take only one course has a choice between this and computer science. Wouldn't the computer course be more valuable? If you had a choice between this and a statistics course, wouldn't the statistics course be more valuable?

Smith: Students are going to take those courses that they are required to take, and what I see is more and more state kinds of requirements. Various states seem to be saying, "We expect these kinds of things from our students that we didn't expect ten years ago."

Curnutt: You've got a lot of students who are required to take one more math class, whatever program they are in, and they don't care what math class it is. What we often give them is intermediate algebra; what a terrible thing to do to anybody. There's got to be an alternative.

In other sciences, there are appreciation courses. Although we don't typically call astronomy an appreciation course, that's what it is. There's a music appreciation course. Mathematics is as good a liberal art as any of those other courses; why don't we have the same kind of alternative?

I really think in the 60's and 70's we were way too ambitious with what we thought we could do with this class. It seemed to me that the intent of the class was, "Hey, let's show these people what mathematicians really do," and I say, "Fat chance." Should we show them what abstract algebra is and what topology is? We haven't got a prayer to accomplish that kind of thing; that's just way too ambitious.

Fusaro: I once taught a math appreciation course. I called it "Environmental Math," and most mathematicians did not think it was mathematics at all. Let me tell you what the topics were. The mainstay of the course was diagrams, developed by H. T. Odom. They are pictorial and from them you can set up what he calls "flow equations," and if you look at it, they're differential equations in disguise, but that's not mentioned. Then you have the students set up execution tracers for them. None of this is by itself even considered mathematical skills, but it's a technique where you are actually solving step-by-step numerical integration. You have them chart it and then, of course, you have them code it. You show them a simple code; you don't tell them they are programming; just code it. So, if you take a little broader view of what mathematics is, this is okay. If you don't, you say, "Well, that's a lot of hooey. You know none of that stuff is really mathematics. It doesn't require algebra, and it has no prerequisites." Students might just as well take any other course on the college campus. I think all this stuff is mathematical. They end up with five models; they have the verbal description, the diagrammatic expression for it, the flow equation, the chart, and the coding.

Statistics--For Whom?

Davis: At a recent workshop we had a group of about one-third industry people and two-thirds academic people. During a quiet moment when we were discussing educational philosophies, I raised the following question: "If you were able to change something in the educational curriculum that you went through, what would be the first thing that you would want to change?" Repeatedly, the industry people said, "Statistics is the most important of the areas of emphasis. It's a subject that should be studied by all, and you should downplay the emphasis on probability and accentuate the statistics."

Watkins: I think that I became converted to statistics as a result of phone calls. You know, mathematicians receive phone calls from the community asking for help with problems. They were all about statistics, except for one in calculus.

Page: There's only so much room in the curriculum. More statistics means offering it at the expense of not offering something else. There's another point of view, though, that I don't see here and that I would like to raise. We agree that statistics is necessary, and it may be possible that many students don't have the opportunity to take a course in statistics. But what you never hear is why aren't statistical concepts integrated into the mathematics curriculum per se?

Watkins: I would like to see the calculus emphasis as the basic core in mathematics courses replaced with the statistics emphasis. One thing that would do for us is that the students who drop out along the way before they get to calculus will have something useful. A little bit of statistics is better than no statistics. A little bit of manipulating rational expressions is worse than none.

Maurer: Ann, in your paper, you've made the nicest one- or two-page description of statistics that I've seen. What is your sense of how much the mathematics community understands that there's been a change in the presentation of statistics? I think everybody has now heard the phrase "exploratory data analysis," but they don't know what it means. Secondly, to what extent does the pre-collegiate mathematics teaching community understand this, because I think what is beautiful about some of these new statistics is that those ideas can be put in with examples at a very early point?

Watkins: They're going to be understanding it soon. This summer (1984) in Princeton, the Woodrow Wilson Foundation is putting on a month-long seminar for high school teachers in the new statistics including exploratory methods. These forty teachers are expected to go out and teach other teachers. What we are doing with NCTM and ASA in December is also along that line. There will be booklets published--we don't know by whom yet--but the Joint Committee of ASA and NCTM, is writing them. I think it's coming, and high-school teachers are becoming more and more aware, and they are asking for materials.

Rodi: I think though that this change in statistics goes against what Warren was saying in the sense that as statistics becomes less mathematical in the traditional sense, it may be harder to put it in.

Gordon: Don, why are you listing statistics on the endangered species list, unless the idea goes back to a comment that Ben made in his paper yesterday that mathematicians have somehow made statistics the pariah of the mathematical community?

Albers: Our CBMS survey data showed the following: from 1975-80, mathematics enrollments in two-year colleges increased by 20%; over the same period, statistics enrollments increased by only 1%. Equally distressing is the fact that statistics accounts for only 3% of total mathematics enrollments.

Tucker: What has happened is that statistics has gotten squeezed out between core math and computer science. Evidence of this could be found at a recent meeting of the Mathematics Advisory Committee of the College Board, which has as its long-term project, "Project EQuality." They are interested in the middle fifty percent of the students; knock off the top and bottom twenty-five percent. What's right for them? Many of them are going on to college, at least to community colleges. Many of them take the College Boards. What questions should be on the College Boards for this middle fifty percent? Recommendations were made--there was also talk about computing, data analysis, and so on--and the consensus of some of us was that students still need the traditional core of two years of algebra and one year of geometry. When one looked at their sample curricula, it was two years of algebra and one year of geometry. Computing was going to go into the junior high school or fourth year math course parallel to calculus. Statistics--what little had been in the curriculum was out--it was going to be piggybacked on top of computer programming off in the corner. There was just enough concern about bringing a little computing into algebra that the probability and statistics went. This is a national phenomenon; graduate schools in statistics don't get the good applicants they got ten years ago because students are all going into computer science.

Ellis: The only time I ever taught statistics was in my first year of teaching. Coming in as a new instructor, the schedules had already been prepared, so I had to teach it. I enjoyed it; I taught two sections of it back to back, worked like crazy, and have not taught it since. It's not because I didn't want to teach it again; it was a pretty good experience, but well, it's just a little uncomfortable. What happens to statistics for engineers and physicists? Are we teaching any of that? The course I taught was for business majors. The engineers need it too, and we need to announce that.

Watkins: One of the problems with statistics that several of you have mentioned, I think without meaning to, is that statistics is often suggested for the students who are only going to take one math course. Somebody else said they suggested it for the students who didn't want hard computational skills. We understand it's important, but who do we steer into it--the very weakest students. I don't understand why we don't tell our very best students, "This is a course you need." When we talk about students in engineering, we are talking about the mathematically talented. We never advise them to take statistics classes. This gives the course a low position in two-year colleges.

Maurer: I think that statistics is now in a peculiar position. Intermediate courses in statistics are easy for mathematicians to teach, and the first course is hard to teach. That is, intermediate courses are still the mathematical statistics courses. If you read the book and if you know some probability, you can teach it. But all those other questions now come in the introductory course, and if you want to teach it well, you can't just read a book.

Page: Ann, one of the statements in your paper might be a little bit misleading: "A 1977 survey of about ten percent of the two-year college mathematics teachers in the United States found that 74% of those holding a Doctorate felt secure about their qualifications to teach statistics, but only 46% of those with a Master's plus one year, and only 42% of those with a Master's felt this competent." My question then is: Is that really per se for statistics, or isn't that true for any course that one has not had and therefore feels a little anxiety about teaching?

Albers: I think that reflects part of your training, Warren. A lot of those doctorates that we surveyed were Doctorates in Education, who usually take some kind of educational statistics, and I think that enhances their confidence, at least, to teach it.

Part 4.

NEW CURRICULA AND NEW TOOLS

IT'S GOING TO HAPPEN ANYWAY...

by

Ben Fusaro

SUMMARY

The first electronic digital computer was invented 42 years ago by a math and physics teacher who was motivated by the wish to solve a system of equations arising from a problem in scienc The Bourbaki movement came to life a few years later, motivated by the wish to emphasize the logical and structural aspects of mathematics. Mathematicians were captivated by Bourbaki, but were not so taken by the computer. Yet, the computer began a transformation of society that the microcomputer will accelerat Mathematics has been affected by the computer revolution, and will go through profound changes. *Mathematicians* can continue to engage in timid or delayed interaction, or they can take leadership roles.

A historical and philosophical sketch is used here to suggest a conceptual framework for an emerging symbiosis of mathematics and the microcomputer. Two-year colleges have been typically more able than four-year colleges or universities to strike out in new directions. Here is a golden opportunity for two-year college math (science) departments to put this flexibility to good use, to go beyond responding to a current need.

CONTENTS

IT'S GOING TO HAPPEN ANYWAY ...

B.A. Fusaro

1. *HOW IT BEGAN* In the mid-1930's, a sequence of events was unfolding at Iowa State College that were to culminate in a remarkable invention. A professor of mathematics and physics was studying the polarizability of helium in an electric field. The model for this experiment turned out to be a wave equation. As usual for such partial differential equations, the only feasible method of solution was numerical. This gave rise to a fundamental problem, that of solving a system of linear algebraic equations. The challenge of solving such systems with something faster than the available mechanical devices became his Archimedean problem. He became obsessed with designing a machine. Two years later, in 1938, he had his *Eureka* experience. Four basic concepts for the modern computer had crystalized in the mind of one John V. Atanasoff. He would build a *binary*, *electronic* machine that used *logic systems* (not enumeration) for processing, and *automatically refreshed* capacitors for memory units. Atanasoff's fertile brain, complemented by his own skillful hand and the assistance of graduate student Clifford Berry, built the world's first electronic, digital computer -- the *ABC*. This precocious desk-size bantling was unveiled in October 1942. The ABC could solve up to 29 equations in 29 unknowns. It was produced on a miniscule budget, and at a locus far removed from the giants of computing and research -- in an Iowa State College physics lab.

Soon after, what today appear to be electronic dinosaurs roamed the land. It began with the ENIAC in 1944, evolved into the early UNIVAC's, and climaxed in the potent, time-sharing IBM Systems 360 in 1964. The 360 (later, 370) became an industry standard.

Ten years later, the tiny INTEL 8080 four-bit chip was to power the small mammals that would challenge the outsized lizards. MITS, a small company in Albuquerque, N.M. produced the Altair microcomputer, a $500 kit based on the 8080 chip. By 1975, over 600 Altairs had been sold, mainly to hobbyists. In 1976, Jobs and Wozniak built their first Apple Computer, and the rest is history *au courant*.

And where was established mathematics during all of this? Mainly, on the sidelines. Established mathematicians responded to the challenge with the alacrity of a garden slug. The 1981 CBMS report, *Undergraduate Mathematical Sciences* [10] has some trenchant observations on changes since 1975: "Access to computers is up sharply, but the impact of computers has changed little ...." "Computer course enrollments have exploded and outnumber those in calculus."

In 1973-83, M.I.T. showed a decline in math undergraduates from 18% to 6%, while engineering went from 38% to 75%, with 35% of the undergraduates majoring in EE/CS. Many other statistical tidbits about academe could be cited.

The general public is also responding, with microcomputer sales estimated at ten million in 1984. There are likely to be 20 million sold in 1985. Microcomputers have fanned out from their toy-hobby status. A major psychological barrier was breached three years ago with the appearance of the IBM PC -- it was now OK for serious business types to own microcomputers. Just recently, IBM came out with a 32-bit desk-top System 370 at about one-fiftieth the cost and size of its progenitor.

It is too late for mathematics to play its traditional role of the pacesetter in these new developments. However, the microcomputer does give us a second chance to join the parade [3]. The microcomputer has emerged as a significant new force -- it is going to transform society with us, or without us. The engineers recognize that microcomputers are important. Students and the general public know they are important. Do we ...? The survival of mathematics as a healthy, viable discipline depends on the answer to this question. We can go the way of the classics and philosophy, or we can work with the mighty midget and grow.

2. *SOME IRONIES* About the time the computer surfaced, specialization was becoming a way of life for scientists and (especially) mathematicians. Yet, Atanasoff was a generalist. He had a BS in electrical engineering from the University of Florida, an MS in mathematics from Iowa State, and a PhD in theoretical physics from the University of Wisconsin. (Even today, at 81 years old, Dr. Atanasoff urges students who are interested in computer science to study all the mathematics they can.)

The microcomputer revolution, with the MITS opening shots in 1974, has been mainly the province of free-spirited, independen minded young men. These attributes are an important part of th traditional stereotype of a creative mathematician.

The third irony is the most important for the purposes of this paper. In 1942, in Ames, Iowa, Atanasoff invented what might be considered a mathematician's dream -- a device for fas accurate calculation. Its creation was motivated by a problem that originated in natural science. It had the potential to transform mathematics education and practice via an emphasis on the *computational content* of mathematics. Two years later, in Nancy, France, a group busied itself with a quite different activity, something that G.H. Hardy would have considered a dream Its creation was motivated by a wish to clarify, purify and unify mathematics. It had the potential to transform mathematics education and practice via a stress on the syntax and structure of mathematics. This conglomerate product of elite French thought was known as *Bourbaki*.

Bourbaki soon jumped the Atlantic and exerted a powerful influence on established American mathematics. This led to over 25 years of narrowness that historians will view as an anomaly. They will wonder why a supposedly pragmatic people preferred Bourbaki almost to the exclusion of Atanasoff.

3. *THE BOURBAKI YEARS: 1945-1975* Bourbaki influence rose very rapidly in the major universities and then filtered out an down to the rest of the mathematics community. During the thre decades after World War II, established mathematics had become dangerously imbalanced. The overstressing of abstract formalis coupled with an exclusive mentality, alienated our client-departments, aggravated employment problems, and left us in splendidly pure isolation. Classical applied mathematicians

had to take refuge in institutes, or work out of closets. The computer infant faced indifference or aversion and so went in the direction of the more friendly business and engineering departments. Statistics remained in usual quarantine. Operations Research took root in management and industrial engineering.

Service courses slipped out of our hands, while engineers and others surreptitiously, or even openly, taught math courses. (A 1981 CBMS report [10] on two-year colleges notes that since 1970 math courses taught outside of math departments have *tripled*.) Bachelors and Masters degrees tended to be viewed as way stations on the path to a pure PhD.

In 1970, the year that young people first warned the nation that it was in environmental trouble, the employment figures gave mathematics explicit warning of impending tribulations. PhD's in the most rarified of fields were being turned out by the hundreds, with no takers. Many were as pure as the vestal virgins, unsullied by any exposure to computer science, operations research, statistics, or even classical applied math. Did no one smell the smoke or sound the alarm ...?

In 1965, Morris Kline assailed the *New Math*, a K-12 offspring of Bourbaki, as a peril to U.S. scientific progress [5]. He has continued the assault on this symptom of imbalance in academic mathematics.

In 1966, there were two symposia on the Impact of Computing on Undergraduate Mathematics Instruction [9]. A. Ralston gave this summary: "The only remarkable thing about the impact, thus far, of computing on the undergraduate mathematics instruction, is its absence." (A *1981* CBMS report has this to say about computer usage in math courses in two-year colleges: "The small impact of computers in mathematics teaching can be seen by noting that less than 2% of all sections of mathematics reported the use of computer assignments for students." [10, p. 88].) Ralston makes an equally interesting observation about panelist N. Macon's comment on employability: "Macon makes the *controversial* prediction that the current rate of increase of mathematics undergraduates may result in an oversupply of bachelor level mathematicians by the mid-1970's." The italics were added to emphasize that even the outspoken Ralston thought such a sensible prediction was controversial. If one turns to the report on the symposia [9, p. 670], the following remark is found:

"Three panelists and several audience members took exception to the inference in Macon's talk that an oversupply of mathematicians might be produced by the mid-1970's if present growth rate for mathematics majors continue." It was lucky that Macon was not so prescient or bold as to predict an oversupply of MA's or PhD's -- he might have been hooted off the panel. This was the Bourbaki euphoria that permeated mathematics in the 1960's.

In 1967, J. Kemeny predicted that learning to use a computer would be as important as learning to read and write.

The polemics or warnings gradually built up, peaking in 1971-73. An article by Gaskell and Klamkin in 1974 documented the shortcomings of the modern curriculum [4]. They decry the lack of attention paid to the problem, recognition, formulation and follow-up. Their article made it embarrassingly clear that math graduates at all levels had to be retrained in computer science, engineering or statistics before they were of any value to industry. They noted that "Modern trends have even narrowed the field of vision, which may be necessary for true research efforts, but is a handicap for the vast majority of our students who must compete with broadly trained individuals from other fields."

An interesting article written by the current MAA director has a devastating metaphorical title: *England was lost on the playing fields of Eaton: A parable for mathematics* [11].

The job situation was so bad in the 1970's that one report in the AMS *Notices* slyly dropped non-PhD's from the unemployment figures. Chairmen dreaded the unsolicited letters from young, trusting mathematicians, who had no forewarning of what was in store for them. The PhD candidates were naive, but not fools. As the market feedback made its appearance, students began to jump ship.

How did mathematics get itself in such a fix? As the above comments show, it had cut itself off from its applied component. It had a bad case of Bourbaki-itis. Even now, the responses we observe seem to be on an *ad hoc* basis. Something like: " A dash of applied math will make our degrees more applicable, so let's give it to them." I would suggest that no counterpart of the medical needle, nostrum or knife will cure the disease. If we are going to do any more than treat the symptoms, we must turn to *regimen*. And before regimen is begun, we must seek the causes, which means to look at our roots. What led mathematicians down the narrow road whereby abstract, formalist disquisitions -- *pure math* -- became the only value yardstick by which mathematical activity is measured? And what can be done about it...?

4. *THE ROOTS OF MATHEMATICS* All mathematics (for the purpose of this paper) will be divided into three parts: That of Euclid-Plato, that of Archimedes, and that of Pythagoras. The easiest to trace, cleanest to define, and most influencial is the first.

The Euclid-Plato school is the direct antecedent of what is known as pure mathematics. It has been refined, but has not changed in any essential way. It begins with a substrate of logical axioms and immediately proceeds to more "mathematical" assumptions or *postulates*, a small set of related statements that involve a few undefined terms or *primitives*. The logic and strict rules of inference yield further statements, or *theorems*. The creation and study of such *Formal Deductive Systems* (FDS) is the quintessence of pure mathematics. It is not necessary to give an interpretation to the primitives, and is usually considered undesirable. It was rather difficult *not* to give an interpretation of the first and paradigmatic FDS, Euclidean geometry. (There are some scoffers who think that the content was there because of experience and the abstraction came later.) In any case, later systems, such as the non-Euclidean geometries indicate that it is legitimate to consider the FDS as an abstract, idealistic, crystal-like product of the human mind, seemingly independent of sense data. It and its philosophical counterparts were kept alive by scholastics and thinkers through the ages. When we reach Newton (born 1642), we see the FDS applied to natural science. The primitives are interpreted, which is to say they are given meaning - as in *mass* or *length* - and the postulates become true or false, as do all of the theorems. Thus, the FDS serves as a model, and these most sophisticated of all models are known as *theories*. (Galileo, who died in 1642, had been the first to work with *algebraic* models, but they were equations, not FDS's).

Logical rigor began to wane as the analysis based on the Leibniz - Newton calculus took hold, and mathematics drifted away from its pure state. Cauchy, Weierstrass and others came in as a cleaning crew, and then it was on to the pure and phantasmal creations of Cantor (1879). A revitalization of the foundations occurred at the turn of the century, at the hands of four mathematical logicians: G. Frege, G. Peano, A.N. Whitehead and B. Russell. It culminated in the monumental Russell and Whitehead "Principia Mathematica" (1910), which purported to reduce all of mathematics to logic. The last important actor in this drama was David Hilbert, who characterized mathematics as a game played with meaningless signs and subject only to the constraints of consistency. Peano and Hilbert's formalistics are lumped with the others' logistic, since the differences are much less important than what they have as a common foundation, the Formal Deductive System.

And so, on to Bourbaki. This school of mathematics, still dominant today, has the following attributes: Its objects of play are abstract, formal, ideal. It is concerned with syntax and structure. And (to protect Hilbert from a great man's indiscretion) the primitives are uninterpreted. Semantics, content, truth or falsity -- all are irrelevant. As for existence, *it stands for nothing more than consistency.* Even the self-consciously pure mathematicians, such as G.H. Hardy, do not always deal with Formal Deductive Systems. However, the FDS is their paradigm, and they maintain a high level of logic with no interpretations as crutches. This (almost) completes the story of the Euclid-Plato school.

The Archimedean school can be thought of as a point of view that began with Archimedes, the father of (classical) applied mathematics. He was familiar with FDS's, but chose not to restrict himself to them. He used graphics, inferential arguments, and (probably) physical models. Kline has referred to him as the greatest mathematician of antiquity. His approach to mathematics has been used through the centuries by scientists and mathematicians, among them the great Euler. A repudiation of such relatively informal methods began with Lagrange (1797), who declared his independence of graphs. Many engineers, physicists, applied mathematicians, and others employ Archimedes' mixed methods.

The Pythagorean school was well defined in antiquity, but its trace through history is fuzzy. Pythagoras' outlook went beyon

Archimedes' in that he wished to apply mathematics to both the natural external world and one's internal (spiritual) world. Thus, Pythagoras' view of mathematics had a subjective-mystical content that was absent from the other two schools. On the other hand, it had a stringent view of acceptable mathematics. First, there was an insistence on construction and second, the mathematics was restricted to integers or ratios of integers. (The discovery of irrationals was seen as a threat...)

If one journeys in time to the late 1800's, to the world of L. Kronecker and L.E.J. Brouwer, it is perhaps not too far-fetched to see Intuitionism as a distant relative of the Pythagorean School. The Intuitionists had a subjective-mystical component that was not shared by the classical practitioners or the Cauchy-Weierstrass clean-up crew. On the other hand, they were logically more restrictive. For one thing, they rejected the law of the excluded middle. Moreover, they insisted that integers were the ground state of mathematical reality and were scandalized by the free-wheeling antics of Cantor.

Moving into the 1960's, we find the Constructive Math of Errett Bishop. Although the mysticism is gone, a subjective aspect remains. In fact, Bishop claimed there was no way to capture the content and meaning of mathematics via formalistics. Logic is employed at the common (mathematical) sense level, which is learned by *doing*. However, proofs are *more* demanding than in the other two schools. This is due to a critical difference in the concept of *mathematical existence*. The usual view is *existence* means *consistency*, or freedom from contradiction. For Bishop *there exists B* means that *you can construct B*. Thus, indirect proofs, or proofs by contradiction are not acceptable when the hypothesis asserts the existence of something. Such efforts might be looked on as a pointer, or a way to increase one's degree of confidence that a proof -- *or construction* -- might be discovered. Since the constructivist approach divorces many cherished results from theoremhood, it is easy to see why it is not very popular.

5. *THE MATHEMATICAL SCIENCES* J.W. Kenelly of Clemson University was perhaps first to use the designation, in 1971, *Department of Mathematical Sciences*. The term is unwieldy and somewhat inaccurate -- since it does not mean that mathematics

has switched from an inferential to an empirical discipline. But it was (and is) an important signal that the prevalent narrow confines were being rejected in favor of a broader, eclectic view of the field.

In 1974, I suggested a tetrahedron model for the mathematical sciences concept. At the apex is Pure mathematics above a base of (classical) Applied mathematics, Statistics, and Computing. (In my original presentation C denoted *computer science*, since I was imbued with the naive hope that mathematics departments would recognize its mathematical content and adopt the teen-ager as a member of the family.)

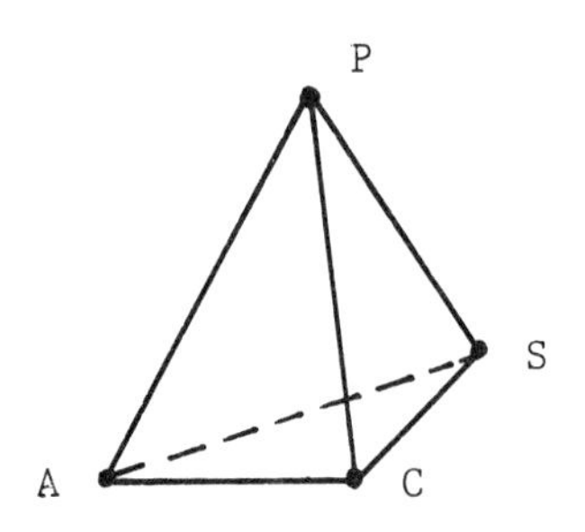

Different subfields can be represented by various regions. Thus, probability might be represented by a band running from P to S, while operations research might be indicated by a strip going from A to S. The apex is sharp and clean. The base makes a fuzzy merge with the empirical world.

6. *CURRICULUM CLASSIFICATION* The established divisions of *Algebra*, *Analysis*, *Geometry-Topology* (with an occasional fourth one, *Applied*) reflect a specialist's slant, and are useful at the level of university research. Perhaps we in the colleges have been overly influenced by this once useful classification scheme.

Kemeny, Snell and Thompson, with their pioneering "Finite Mathematics" (1956) thrust discrete mathematics, with interesting and important applications, into the freshman offerings. F.S. Roberts' "Discrete Mathematical Models" (1976) appears almost as a two-decade commemorative. Nevertheless, in 1984, there are some who feel the need to keep the discrete at bay. Since computer mathematics is by its nature discrete, it is likely that discrete (finite) mathematics will penetrate the curriculum as a component of most courses and also as a course of its own. I suggest that *continuous* and *discrete* be accepted as fundamental classifications in our curriculum thinking.

7. *THE MATH SCIENCES AND THE THREE SCHOOLS OF MATHEMATICS*

The tetrahedron model in section 5 will be expressed very differently in a curriculum, depending on which school dominates our outlook. Let us return to each school.

The Pure (Euclid-Plato, Bourbaki) approach provides precision, structure and a logical basis. These properties strongly suggest that there is also an accompanying certitude, but here is where we must be careful. Some of the arrogance that is sometimes attributed to mathematicians might have its origins here. Yet, Godel's 1931 result tells us that any Formal Deductive System that is adequate for *even just arithmetic* is either inconsistent or incomplete. So the FDS will be logically flawed or else incapable of proving a statement that (under our interpretation) is known to be true. Godel tells us there are *inherent limits* to what pure mathematics can do. Somehow, we do not seem to be able to accept this result. It almost seems as if the attitude is: "If we don't think about it, maybe it will go away." A look at a parallel situation in physics, bolstered by a trenchant observation of B.B. Mandelbrot might be instructive.

In studying or teaching the history of science and math, one notes that 1879 is surrounded by important happenings. That year marks the death of Maxwell and the birth of Einstein, both men being associated with the shift from action-at-a-distance to the *field* concept. A new physics was definitely on its way. It was turbulent, exciting, and filled with healthy controversy. About the same time, as conventional history would have it, mathematics was witnessing a young, creative, freedom-loving Cantor being harassed by some cranky old-timers. Cantor and the free-spirits won out. So math also went through turbulent, exciting times, but filled with disturbing controversy.

Although there was something awry with turn-of-the century mathematics, I was not able to articulate it. This was not just a question of controversy -- a basic shift had occurred which was difficult to explain. B.B. Mandelbrot gave a brilliant lecture on fractals at the University of Florida (where I was on sabbatical) in 1981, and suddenly the puzzle fell in place. The gist of his remarks appear in his "Fractals and the Geometry of Nature", 1982 [7]. After commenting on the high degree of complexity exhibited by nature (page 1), Mandelbrot writes, "Mathematicians have disdained this challenge, however, and have increasingly chosen to flee from nature by devising theories unrelated to anything we can see or feel." He then quotes (page 3) F.J. Dyson's summary of the theme underlying fractal research:

> 'Modern mathematics began with Cantor's set theory and Peano's space-filling curve. Historically, the revolution was forced by the discovery of mathematical structures that did not fit the patterns of Euclid and Newton. ... The mathematicians who created the monsters regarded them as important in showing that the world of pure mathematics contains a richness of possibilities going far beyond the simple structures they saw in Nature. Twentieth-century mathematics flowered in the belief that it had transcended completely the limitations imposed by its natural origins. Now, as Mandelbrot points out, Nature has played a joke on the mathematicians. The 19th-century mathematicians may have been lacking in imagination, but Nature was not. The same pathological structures that mathematicians invented to break loose from 19th-century naturalism turned out to be inherent in familiar objects all around us [via fractal analysis].'

The term *humanistic conceit* occurred to me as I heard a similar message from Mandelbrot in the University of Florida auditorium. While the physical scientists were learning from nature, Cantor and modern mathematics were on a narcissistic intellectual trip.

There is a second interesting parallel between physics and math. In 1824, S. Carnot published his gem on heat engines. Physicists were faced with a strange new law, formulated a few years later by R. Clausius as the *second law of thermodynamics*. There were limits set by nature on what an engineer or physicis could do. No matter how clever he was, how ingenious his design, how superior his materials or fuel, any heat engine operating between the temperatures T_{lo} and T_{hi} had an *upper bound* on its efficiency, $(T_{hi} - T_{lo})/T_{hi}$. Scientists had bumped up against a baffling limit, but they accomodated their thinking to it. About a hundred years later, they came across an even more baffling limit, W.Heisenberg's Uncertainty Principle (1927 The observer had an upper bound on the accuracy of paired obser vations (such as displacement and velocity). As one became more accurate, the other became less accurate. The startled physicists recovered, absorbed this second big lesson from nature, and went on. (It is ironic that about the same time,

Hilbert was heralding the power of formalistics.)

Now contrast this with how mathematics reacted to Godel's 1931 *limiting* (meta) theorem. It was so deep that it took years for even experts to absorb, but by 1945 its conclusions were understandable to professional mathematicians. And what were its effects? Established mathematics went on as if it had never happened -- the FDS was still emperor.

Being an expert in Pure math does not turn one into a god-like creature. Like all other man-made systems, it has its shortcomings and flaws. As depicted on the tetrahedron model, it is simply one component of the mathematical sciences. It might be pedagogically unwise to expose a logical flaw to our young charges. However, if we have a gut-level awareness of this limitation, we might come across to our students and non-math colleagues, as less remote (and less omniscient).

The Archimedean approach, essentially that of the "working" mathematician, seems to fit comfortably in the math sciences scheme. It is at home with real-world applications and is not concerned with foundations or historical controversy.

Now for the third, distantly Pythagorean school. It has no explicit locus on the tetrahedron, and yet it is of great importance because of the way it can modulate how we teach mathematics.

It's entree is through vertex C, the computing or computational aspects. Henry Cheng has addressed this point in his *Constructive mathematics and computer science*, 1972 [2]. He notes that one of the main aims of Bishop's Constructivism is to inject the theorems of conventional math with computational content. In addition, the constructive approach restricts itself to finitely performable sequences of operations. This algorithmic approach allows a tie-in with an ALGOL (or other high level language) program. Thus the algorithm is a link between computer science and mathematics.

At every level and in every course, the question needs to be asked: Are we emphasizing computational aspects? Have we explored the *computational* aspects of the method or theorem? If it is an indirect proof, is there a constructive proof available? Are we choosing errors, bounds or tolerances as reals, when they could simply be integers or reciprocals of integers?

8. *THE COMPUTER AND THE CURRICULUM* The two-year colleges serve a significant number of students in higher education, approximately one-third of all college enrollments. Many of the students who enter college this fall will graduate in 1986 with an AA degree in hand. In that year approximately *30 milli* microcomputers will be sold. Vocational-technical, career, or liberal arts students -- it does not matter -- they will face a profoundly different society from their counterparts of just two years ago. The electronic marvels of their world will oper ate on principles very much like the 1942 desk-size *ABC*, plus a stored program. But there will be no punched-card input/outpu only the familiar keyboard, printer and high-resolution screen. It will have massive external storage, about a million characters on each pocket-size disk, and an equal amount of internal memory. It will be able to do any of the scientific computing, word processing, and perhaps even *symbolic* manipulation of machines that in the 1970's cost up to $50,000. In fact, the electronic marvels of the 1970's will appear like bulky museum pieces to the younger graduates.

The AA's who transfer to engineering schools will find that scores of them will require all students to have their own micr computers. The graduating AA's will be on the doorstep of an information-based society. As W.F. Lucas of Cornell has pointe out, "The impact of the computer is much, much bigger than most of us can conceive." [6]

There is an emerging symbiosis of *mathematics* and computer science. To date *mathematicians* have only been marginally involved. The microcomputer now gives us a second chance. It will not do to merely *cope* with the computer, or view it as a nuisance that must be contained. Just as a text, dictionary or thesaurus is a tool for augmenting the reach of the mind, so to is the microcomputer such a tool, a dynamic, interactive one.

Plans need to be made now for making the computer a part of *every* math sciences course. Every course in the kind of mathematics envisaged here would have a graphical or computational component. If the formal aspects of the subject, such as geometry (or topology) are not of a numerical nature, the intuitive aspects can be strengthened by graphing. We will be able to *do* some very important things that we now merely talk about: Numerical integration and composition of curves (Calculus), graphing fields of lineal elements and qualitative behavior (D.E.), and operations with realistically large matrices (Linear Algebra). Colleges are responsible for seeing that the 1986 graduate will have an experience in computers. For math sciences students, that minimal course ought to be a programming course that *emphasizes structured design and problem-solving*. If only get-the-code-out-the-door courses are available, math sciences needs to teach its own. (If the department has so removed itself from computing that such a course is not seen as part of the math sciences, it might be better to use only packaged software.)

Departments that have proudly maintained that "...mathematics requires only the human mind plus paper-and-pencil..." need to start singing a different refrain -- quickly. We need to convince administrations that the *math sciences need equipment budgets*. A good argument might be: *Microcomputers cost no more than microscopes. -- Which gives more educational return for the money?* (Administrators, weary of the endless budget-gobbling of biology departments, might respond to this one.)

There are now many one-day to two-week workshops or seminars designed for faculty who want to learn topics in computing. Many are relatively cheap. Currently, only a *small* proportion of those who take advantage of these are mathematicians. However, AMATYC and the MAA have been offering more experiences in computer-related topics each year.

Available software has hitherto been a weak link in efforts to infuse the computer in the math sciences curriculum. Sometimes code (usually unstructured BASIC or FORTRAN) was pasted into standard math texts. However, there are some materials available in which text and software are interwoven. Most of us in teaching do not have the time or expertise to produce the software needed. The authors who *are* producing these classroom-ready materials are playing a vital role in the evolving role of the math sciences.

INTEGRATED SOFTWARE

Following is a short list of works in (pre-)Calculus and Algebra:

R.L. Finney, D.T. Hoffman, J.L. Schwartz, & C.O. Wilde, "The Calculus Toolkit", Addison-Wesley, Reading, Mass. 1984. This is an interactive BASIC software package. It has four diskettes formatted for the Apple II+, IIe. (Diskettes for the IBM PC are in preparation.)

Garry Helzer (Univ. of Maryland), "Applied Linear Algebra with APL", Little Brown, Boston, 1982. (*)

K.E. Iverson (I.P. Sharp, Toronto), "Elementary Analysis", APL Press, Palo Alto, Calif., 1976. (*)

D.L. Orth (IBM, NY), "Calculus in a New Key", APL Press, Palo Alto, Calif., 1976. (*)

C.C. Sims (Rutgers Univ., NJ), "Abstract Algebra - A Computational Approach" John Wiley, N.Y. 1984. (*)

Gareth & Donna Williams (Stetson Univ., FL), *Linear Algebra Computer Companion.* This is an interactive BASIC software package. It is on two diskettes formatted for the Apple II, IIe. Designed to accompany the G. Williams text "Applied Linear Algebra with Applications", Allyn & Bacon, Rockleigh, N.J., 1984.

The following two authors have developed extensive (but not *commercially* available) software:

G.J. Porter (Univ. of Pennsylvania), Computer Graphics for the Apple II.

F. Hickernell & W. Proskurowski (Univ. of Southern California), Calculus for the DEC Gigi.

---

* The software is part of the text in the form of the powerful scratch-pad language, APL.

REFERENCES

1. Errett Bishop, "Foundations of Constructive Analysis", McGraw-Hill, NY, 1967.

2. H. Cheng, *Constructive mathematics and computer science,* Proc. ACM Annual Conference, Boston, August 1972, 986-990.

3. B.A. Fusaro, *The microcomputer - a second chance,* SIAM News, 9/83. Reprinted in MAA *Focus,* 3/84 and *Collegiate Microcomputer,* 3/84.

4. R.E. Gaskell and M.S. Klamkin, *The industrial mathematician views his profession,* MAA Monthly 81 (1974), 699-716.

5. Morris Kline, *The New Math,* NYU Alumni Magazine, August 1965.

6. W.F. Lucas, *Mathematics Curriculum Conference* (L. Steen, Organizer), St. Olaf College, Minnesota, 14-15 November 1980.

7. B.B. Mandelbrot, "Fractals and the Geometry of Nature", Freeman, San Francisco, 1982.

8. A. Ralston, *University courses and programs,* ACM Computing Reviews, Nov-Dec 1966.

9. *A Report of Two Symposia on the Impact of Computing on Undergraduate Mathematics Instruction* (Atchison, Givens, Macon, Murray, Goldstine; Duren, Hamblen, Poor, Shanor; Miles, chrmn), Florida State University, 14 March 1966. ACM Comm. 9 (1966), 662-670.

10. *Undergraduate Mathematical Sciences in Universities, Four-year Colleges, and Two-year Colleges, 1980-81,* J.T. Fey, D.J. Albers, W.H. Fleming, C.B. Lindquist, CBMS Report of Survey Comm., IV, 1981.

11. A.B. Willcox, *England was lost on the playing fields of Eaton,* MAA Monthly 80 (1973), 25-40.

(DISCUSSION BEGINS ON P. 289.)

DISCRETE TOPICS IN THE UNDERGRADUATE MATHEMATICS CURRICULUM: HOW BIG A STEP SHOULD WE TAKE ?

by

Sheldon Gordon

SUMMARY

The academic use of computers and the rapid development of Computer Science programs have produced calls for the introduction of discrete mathematics into the undergraduate mathematics curriculum. The present article is intended as an overview of how discrete mathematical topics can be incorporated into the standard mathematics offerings in the two-year college curriculum.

DISCRETE TOPICS IN THE UNDERGRADUATE MATHEMATICS CURRICULUM: HOW BIG A STEP SHOULD WE TAKE?

Introduction.

More than three centuries ago, Newton and Leibniz blazed a new trail in mathematics that led to exciting new vistas on the physical universe. This approach was so immediately successful that every succeeding generation of mathematicians has unquestioningly followed the same path which has led smoothly and continuously to ever new heights and achievements. In the process, the trail has broadened until now it is a twelve lane superhighway covering applications not just in the physical sciences, but also in the biological sciences, the social sciences, business and the management sciences and computer sciences.

Over the last decade, however, there have been some whisperings that the outside lanes of this highway might not provide adequate support to all people using mathematics. It now seems that the very success of the continuous approach may have caused us to overlook other vistas of potentially equal significance and value. In part, this is quite understandable. Prior to the development of the computer, the great difficulty of performing extended numerical computations made it discouraging to treat problems requiring iterative calculations. Not only is this drawback now gone, but there has been a complete reversal whereby discrete calculation is now considerably less labor that calculations with continuous functions. This new capability has now led to considerable discussion and debate (See [3,4]) about the advisability of incorporating varying amounts of so-called "discrete mathematics" into the mathematics curriculum at all levels.

The present paper will attempt to address this question by examining the different types of courses typically offered in the first two years of college and seeing how appropriate it might be to incorporate discrete topics into existing courses in the math curriculum. Appendix I contains a list of those topics which currently go under the heading of discrete mathematics. Throughout, we will consider the problems associated with actually trying to implement such endeavors.

First of all, let's consider the possible reasons for including discrete topics in mathematics offerings. Basically, they seem to come down to the following:

1. for their own sake;
2. for computer applications;
3. for their use in applications in other subject areas;
4. for their ability to interest and motivate students.

It is the last three items that should dictate to what extent, if any, such curricular changes make sense. The first is primarily an individual judgment.

To gain some perspective, it is a good idea to consider briefly how non-mathematicians view discrete mathematics. Physicists will tell you that the universe is inherently discrete (quantized) and that a continuum exists only in a mathematician's imagination. Of course, the physicist then goes blithely ahead and uses continuous mathematical models because they work so well. On the other hand, Greenspan [2] has been able to derive virtually all important physical concepts and formulas based solely on discrete formulations involving purely arithmetic and algebraic operations. Further, engineers have realized for years that there are situations which are inherently discrete and so require discrete mathematical methods, the so-called discrete time or sampled data systems. Accordingly, they have developed a wide variety of sophisticated discrete mathematical techniques to handle them. Closer to home, statisticians have dealt with both continuous and discrete methods virtually on a par and have developed parallel mathematical structures for both types of situations.

The Calculus Sequence.

Since the original impetus for continuous mathematics began with the development of calculus, it is only fair to use the same starting point for a discussion of the role of discrete mathematics in the undergraduate curriculum. Several years ago, with support from the National Science Foundation, I developed a new approach to calculus which I called a Discrete Approach [1]. The idea is to incorporate many topics from the calculus of finite differences and sums into the traditional calculus sequence. The rationale for this is:

1. The finite calculus contains discrete parallels to most of the important concepts and methods in calculus. However, since the finite calculus is based on the concept of sequence, its development and applications can be handled using only arithmetic and algebraic operations. This makes the ideas more accessible to weaker students.

2. Once a discrete topic has been treated, the corresponding continuous result can be obtained by resorting to the limit as the step (or increment) approaches 0. Thus, each idea is essentially treated twice. This constant review makes for better student comprehension of the concepts.

3. The non-calculus nature of the development allows solid applications such as discrete max-min analysis and growth and decay problems to be treated within the first week of the course. This provides tremendous motivation to students whose interests are in the use of mathematics, not in the mathematics itself.

4. The discrete setting provides the mathematics necessary to utilize a computer as an integral part of the course - after all, any computer application involves discretizing a function.

To illustrate this discrete approach to calculus, I will describe one of the topics that I cover about one week into the course: finding the approximate location of maxima and minima for a function. By this point, the students have seen the concept of the forward difference operator for a function,

$$\Delta f(x) = f(x+h) - f(x),$$

for any step h. It is then quite apparent to them that when the difference is positive, the function increases (or at least experiences a net overall increase across that subinterval) and when the difference is negative, the function decreases. Further, when the difference is nearly zero, the function is close to its maximum or minimum value. Therefore, the students are able to preview one of the most powerful applications of calculus long before any of the essential machinery is developed and even before being introduced to the concepts of limit and derivative. Further, it is easy to develop the notion of taking smaller values for the step to get a better idea of the location of such extrema, again previewing an essential element of calculus. Moreover, this approach is a very natural one to implement on a computer using a simple search method.

The results of this project were extremely successful both with my own and various colleagues' classes. Student performance on exams was at least as good. Interest and motivation were considerably higher than usual. In fact, the use of the finite calculus made things so much easier to the students that it was possible to move faster than usual through the course. Thus, there was virtually no loss in other topics to make room for these.

There are a number of very important lessons from my experience with this approach and its implementation which I think we should all bear in mind when considering significant changes in the math curriculum. First of all, mathematicians tend to be a very conservative breed, especially when it comes to traditional courses. Further, at many schools, the standard courses such as calculus are offered primarily for the benefit of other departments, such as physics and engineering, and to an extent, we have to structure the courses to their specifications. Finally, courses at two year colleges have to be structured with a view to transferability to four year institutions and this imposes an additional restraint against changes. Together, these factors make for an almost overwhelming inertia that keeps things going in a fixed course. The fact that a better method or approach may exist does not necessarily mean that it will be widely adopted and implemented.

Consequently, if we want to introduce any significant changes in mathematics curricula, then we should focus our attention and efforts on non-traditional math offerings. For example, I believe that many schools are on the verge of developing a new calculus track for computer science majors -

there is too much dissatisfaction with the traditional calculus sequence in terms of its relevence to the computer science curriculum. The overwhelming success of calculus in addressing and solving problems from so many areas mitigate against dropping calculus altogether as a requirement for computer science students. On the other hand, virtually all the mathematical work that arises in computer science is basically discrete in nature - continuous problems are simply discretized to allow the computer to handle them. Therefore, it is reasonable to expect that a compromise will involve the development of a new track that merges a certain amount of discrete parallels into the standard calculus so that the computer science students have an appropriate context in which to use calculus. Further, such a course is likely to use a more intuitive approach based on the sequential notion of limits, difference and sums. This would provide an ideal context in which to develop some of the mathematics (primarily numerical methods) necessary to implement the use of the computer to solve calculus-type problems. It would also demonstrate the connection between calculus and computer science. As such, it would be an effective blend of continuous and discrete mathematics.

Applied Calculus.

Similarly, arguments can be made for the inclusion of other topics of a discrete nature in the various applied calculus courses that are offered to business, social and biological science majors. These are areas where traditional calculus-based mathematics is imposed on situations that are inherently discrete and which at best are approximated by continuous models. For example, discrete and continuous predator-prey models in biology produce different results - which one is based on a more reasonable set of assumptions? Marginal analysis in economics is based on the incremental changes associated with a unit increase in sales or production. Is it really appropriate to resort to a limit? In fact, in these and other comparable situations, discrete situations are modeled by continuous processes which, in turn, are solved in practice on computers by discretizing the mathematics. At best, this represents circular reasoning; at worst, it leads to errors in the final results.

Would it not be simpler and more logical to introduce some of the comparable discrete models in the courses for these students? After all, topics such as sequences, recursive relations, and solutions of difference equations are considerably simpler than many of the concepts of calculus and are potentially of greater use. For example, growth and decay problems can be treated in terms of the simple difference equation $A_{n+1} = k A_n$ which leads quickly to the solution $A_n = A_o k^n$. This is far simpler and is more meaningful in terms of the usual applications of such problems than the standard approach involving integration and properties of logarithmic and exponential functions. For that matter, how many banks compound interest continuously?

Differential Equations and Linear Algebra.

Just as the calculus of finite differences and sums is an almost exact mirror image of the usual continuous calculus, the theory and concepts of difference equations are parallel to those of differential equations. As such, there is much that could be gained from a course that encompasses both. This is particularly true of those differential equations courses which include segments of linear algebra, since the combination of difference and differential equations would provide a pair of complementary systems to which the algebra is applied. Further, the opportunity to experience both continuous and discrete models in a single course would be a valuable opportunity for the students. For instance, the classical discrete and continuous predator-prey models could be studied and the results of each compared.

With this approach, the students would thereafter be in a much better position to select between appropriate discrete and continuous models to find the one that better describes a particular application or problem that might arise in other courses or in their careers. As things stand now, there is an almost total bias towards using purely continous models. (A similar bias is created in most math students towards the use of deterministic models over stochastic models.) Finally, the availability of difference equation notions and the underlying discrete structures involving sequences and recursion relations would be a major advantage in introducing numerical procedures into such a course in a coherent and natural manner.

Despite all of these advantages, I do not see such a course evolving in the immediate future. If nothing else, the present versions of elementary differential equations courses are simply too heavy to leave room for additional material, no matter how good the justification. This applies to traditional equation-solving courses, theoretically oriented courses emphasizing existence and uniqueness results, computational and computer oriented courses, and those having a linear algebra component. Moreover, the factors dicussed in the previous section on the Calculus sequence certainly apply here as well.

Finite Mathematics

Of all the standard courses offered at the undergraduate level, the one that has come closest to containing significant amounts of discrete mathematics is the finite mathematics course. In fact, the primary difference between finite mathematics and the new discrete mathematics courses is that the former emphasizes applications towards business and related fields. On the other hand, the new courses in discrete mathematics are directed primarily towards the mathematical needs of computer science students and so tend to be somewhat more sophisticated mathematically and to have a greater variety of mathematical topics.

The one change that is likely to occur in the finite math course over the next few years involves an increased use of the computer with particular emphasis on software to handle matrix manipulations. Once the students can have all matrix operations performed for them by machine (or conceivably by a future version of the hand-held calculator), it is not very sensible to force them to perform such operations by hand. Thus, I foresee a reduction in the time spent on matrix operations and related methods such as row reduction operations. The time saved is likely to be devoted to including other topics of a discrete nature. Incidentally, while I do not feel that students in finite math courses should be required to perform matrix manipulations when appropriate software is available, it is still an essential ability for students in linear algebra.

Statistics.

The nature of the subject matter in the standard non-calculus based elementary statistics course already imposes a considerable emphasis on topics from discrete mathematics. In addition to the discrete distributions such as the binomial, hypergeometric and Poisson, many subsidiary discrete topics such as set theory, counting principles along with permutations and combinations, tree diagrams and sample spaces are integral parts of the course. There is really little justification for including additional topics in discrete mathematics in such a course.

However, it is reasonable to expect that many of these discrete topics will increasingly find their way into the precollegiate mathematics curriculum. Therefore, we can expect to see more students coming into elementary statistics courses who have had a previous exposure to some of the topics we normally cover. Consequently, we should anticipate spending less time on such topics in the future. Additional topics of a purely statistical nature will then have to be added. Alternatively, we might well see some of this extra time devoted to real-life applications of the subject matter as students conduct experiments and surveys and then perform the statistical analyses using simplified statistical software packages. Therefore, we should expect a certain degree of evolution in the content and emphases of such a course over the next few years.

Survey of Mathematics for Liberal Arts Students.

Next to finite mathematics, the traditional survey of math course has probably been the course having the highest discrete mathematical content. This will likely continue. Considering the high interest that the general public and hence many liberal arts students have in computers, it seems reasonable to emphasize them and the mathematics underlying them more heavily in such a course than most of us do today. Thus, set theory, logic, induction, recursion relationships, graph theory, combinatorics, Boolean algebra, probability and the like should remain among the core topics for such courses. Further, emphasis on the algorithmic approach as a vital component of mathematical thought is a natural addition to such a course.

Remedial Mathematics.

Rather than focusing on individual courses in the remedial area, I think that some general comments are in order. There are some significant differences in the reasons such courses are offered in college and in secondary schools. In high school, courses such as algebra and even trigonometry are taken by the "good" students because it is the expected thing to do and they presumably know what their future academic needs will be. By the time people get to college, they take remedial math courses because of specific requirements of a curriculum or the need to remove an academic deficiency. In either event, the math is not taken for "fun", as a rule, but rather to develop very specific skills that are usually required for other courses in mathematics and related (as well as unrelated) fields. We cannot afford to lose sight of this. Very few of the topics covered by the term discrete mathematics really add to the arithmetic, algebraic and trigonometric skills that these students need to master. (The binomial theorem may be one notable exception.) As a result, I see very little potential impact of discrete topics on these courses.

In a related direction, relatively few students taking remedial mathematics ever reach high level math and science courses. Many are content with, or require no more than, an elementary course in statistics, for example, and are therefore well served by an earlier exposure to some discrete topics. By the same token, students enrolled in business programs may require only statistics, finite mathematics and low-level applied calculus. For all such students, it makes sense to eliminate many of the more sophisticated manipulative methods from algebra and trig courses and possibly replace them with some appropriate discrete mathematics. However, this clearly necessitates developing various remedial tracks depending on the eventual goals of the students. Hopefully, most colleges will move in this direction over the next few years and so better serve all their students.

There is another fact about these courses that should be considered. A very high percentage of the students we see in remedial math courses have already tried the course in high school and have either failed or had limited success. Either way, they come into the college courses with certain preconceptions of what algebra or trig is all about and it is usually very difficult to teach them a better or easier or more sophisticated way to deal with problems. (A notable example of this occurs if we expect them to use anything other than the slope-intercept form for the equation of a line.) Further, if we do things too differently, we run the risk of confusing rather than educating such students. This danger can be compounded if we attempt to incorporate additional topics that they "know" are not part of the course.

On the other hand, a good case can certainly be made that an infusion of fresh topics can enliven a class and so motivate students. This might be especially true in the case of arithmetic where topics such as simple combinatorics and elements of probability can be used to reinforce arithmetic skills while making the mathematics more interesting. However, I do feel that such efforts should be approached with caution. While they can make some of the non-obvious applications of mathematics more accessible to weak math students, this should not be done at the expense of the necessary skills they must acquire. Also, students who have trouble mastering simple arithmetic concepts and operations may well find the concepts and verbal problems involved overwhelming.

Such topics and motivation are better suited, in some ways, for inclusion in mathematics curricula at the elementary and junior high school level. There, the teacher has considerably more time and flexibility available to incorporate new material or new approaches. Further, it makes more sense to attempt to motivate and interest students before they lose interest in mathematics, as is too often the case at the junior high school level. Some of the topics in discrete mathematics certainly have that promise.

Of course, the overwhelming difficulty with this idea is that many of the teachers at that level are not well prepared in mathematics and so may not be willing or able to teach such material. In many ways, the question of mathematical preparation of elementary and secondary teachers presents us with our most important challenge over the coming decade and is one which we must all address.

One final point may be worth mentioning here. Within the next ten years, we will very likely be faced with making some very far-reaching decisions on the content of remedial courses and on the needs of the students taking them. The availability of symbolic manipulation software such as muMath and MACSYMA will have a drastic impact on our curricula and the philosophies underlying them. It is not unreasonable to expect, for instance, the development of a hand- held version of such a package that each student will own and use as he or she now uses a calculator!

Geometry.

The one remedial level math course that was not mentioned in the previous section is geometry. This falls into a different category in that it is not intended to develop skills on the part of the student. Rather, the usual reasons given for teaching geometry are that it helps develop thinking skills and geometric intuition and provides the student with the set of geometric facts needed in subsequent math courses. For all of this, the course has been dying, both at the high school and college level. Those of us who have been teaching calculus during the last few years have certainly seen the effects of this - the students no longer have any geometric feel for mathematics nor do they have any background in mathematical reasoning and the nature of proof.

As a consequence, it seems to me that geometry is the one place in the remedial curriculum that most needs an infusion of new material or a new approach to regenerate interest. There have been some fascinating reports lately of the use of computer graphics, including use of LOGO, to address this from one direction. For example, many of the major results of geometry can be "discovered" using computer graphics. Thus, students can be led to generalize results through a study of individual cases. Similarly, the inclusion of some discrete mathematical topics may also spark a major improvement. For example, some of the modern ideas in geometry, including graph theory and networks, would fit in very nicely and motivate student interest.

Technical Mathematics.

Many of the comments made above regarding the inadvisability of incorporating discrete mathematics into algebra courses apply equally well with technical math. These courses are also designed to impart specific skills and knowledge to the students and, as such, few of the discrete topics would add anything to the students' abilities. Therefore, it is very unlikely that discrete mathematics will make any significant contribution to such courses.

Implementing Discrete Mathematics in the Curriculum

To this point, I have addressed how discrete mathematical topics may be incorporated into existing math courses. The national trend, however, seems to be in the direction of instituting a self-standing course in discrete mathematics primarily intended for computer science majors. I wonder, though, if math majors would be best served by such an approach. The course would stand by itself and the student would rarely see how discrete mathematics relates to the rest of the mathematical curriculum.

In many ways, this would be analogous to the way that numerical analysis is presently treated. Unless a math student majors in numerical methods, he or she does not see how the subject can and should be integrated into most math offerings from calculus through linear algebra and differential equations and possibly beyond. This lacking creates a severe blind spot in terms of using modern mathematical developments. I fear that the same will hold true of discrete formulations if they are segregated from the rest of mathematics.

Unfortunately, from this point of view, it is much simpler to introduce a new course at any institution, particularly an elective, than it is to change an existing course syllabus. Whenever you add a new topic, something has to go to make room for it. Invariably, people will fight to keep the existing topics (often, though not always, with good reason). Further, mathematicians as a group are extremely conservative - for example, mathematicians have the lowest rate of computer usage in their courses of all the science disciplines. This type of attitude will contribute to resistance to incorporating new topics and emphases in existing courses.

Another problem in implementing curricular changes involves the attitudes of the publishing industry. Very few companies will chance putting out a book that does not have a ready-made audience. Unfortunately, many people and departments use a textbook to define their courses. This certainly is the easiest route in terms of individual preparation; it may also be the safe route in terms of not making an independent statement or judgment about a course. In either event, most math instructors will be unable to implement desired changes without an undue and needlessly repetitious amount of effort. Just picture the number of different people who have been developing computer software independently for their courses over the last decade. However, that effort has its own built in reward in the sense that the person writing the programs is developing an important skill. Developing a set of course notes to incorporate additional topics is probably more in the nature of drudgery. This lack of appropriate classroom materials to introduce discrete topics could provide a tremendous opportunity for a group such as UMAP/COMAP to fill in with an array of appropriate modules until the publishers see well-defined curricula evolve and become sufficiently widely established.

Conclusions

Continuous mathematics is responsible for an incredibly broad range of successes in both mathematics itself and the disciplines where mathematics is applied. Discrete mathematics has the potential for a comparable impact. However, if it is to have such an impact, then it must be incorporated within the existing structure of mathematical education to become available to the widest possible audience. At the same time, we must bear in mind that such changes should not be made simply for the sake of change and that the teachers who follow our lead may not share our vision, motivation and enthusiasm. The lessons the mathematics community hopefully learned from the forced implementation of the New Math should not be quickly forgotten in a rush to implement new and presumably better pedagogical methods.

References

1. Gordon, S. P. A Discrete Approach to Computer Oriented Calculus, MAA Monthly, 1979.

2. Greenspan, D. The Arithmetic Basis of Special Relativity, Int J Theor Phys, 15, (1976), 557-574.

3. Roberts, Fred S. The Introductory Mathematics Curriculum: Misleading, Outdated and Unfair, Col Math J (to appear).

4. Ralston, A and G. S. Young, The Future of College Mathematics, Springer-Verlag, New York, 1983.

Appendix
Topics in Discrete Mathematics

Set theory
Permutations
Combinations
Binomial theorem and coefficients
Relations
Graph theory
Trees
Mathematical induction
Equivalence relations
Functions

Truth tables

Algorithmic methods

Difference equations and recurrence relations

Partitions

Recursive definition

Propositional logic

Boolean algebra

Probability

Partial orderings

Numeric functions

Summation notation and concepts

Positional notation

Non-decimal bases and base conversion

(DISCUSSION BEGINS ON P. 289.)

CALCULATORS DO MORE THAN COMPUTE

by

Joan R. Leitzel

SUMMARY

This paper describes changes in the approach to algebra and geometry that have resulted from using handheld calculators with remedial students at Ohio State and with twelth grade students who have no skills in elementary algebra.

The approach described is highly numerical and depends on concrete problem situations to suggest key concepts of algebra. Calculators are used to give students greater confidence, to get them past arithmetic difficulties and into significant mathematics more quickly, and to enable the investigation of realistic problems. More importantly, calculators provide an entrance to mathematics quite different from traditional approaches. Several features of this numerical approach are discussed:

-- using numerical methods to solve problems that students are not ready to solve algebraically or that are more appropriately solved numerically.

-- encountering basic concepts and relationships in problem settings before the concepts are formalized.

-- investigating problems in many special cases so that variables can be introduced in a natural way to describe the general case.

-- using calculators to draw students' attention to arithmetic ideas that otherwise would be of little interest to them.

-- using calculators to permit the early introduction of graphing so that graphing can provide a bridge between numerical work and the formalism of algebra.

CALCULATORS DO MORE THAN COMPUTE

Joan R. Leitzel

Over the last several years a great deal of the experimentation in mathematics instruction has been with different types of delivery systems for traditional college mathematics content. We know about Keller plans, modified Keller plans, computer assisted instruction, computer managed instruction, mathematics laboratories, mathematics workshops, peer tutoring, video-tape support, mathematics modules, self-pacing, retest options, and so on. Although many departments feel these developments have been useful in improving their instruction, these developments generally have not questioned the appropriateness of traditional mathematics content, the way basic mathematical ideas are approached, or the order in which topics are sequenced. These are issues that now need attention.

This paper will not attempt to contribute to the discussion of possible significant changes in college-level mathematics content. It will rather share experiences that suggested changes in the way basic mathematical ideas are approached and the order in which topics are sequenced.

At least two phenomena press us to take seriously the need to re-evaluate what we are doing. One is the evidence that large numbers of students are not learning mathematics well in our college preparatory programs. For this reason, as we seek better ways, we must also attempt to understand what causes present programs to be ineffective. A second phenomenon, of course, is the availability of low cost calculators and computers. These we sense should give new ways of communicating mathematics and even open up new topics for our students.

My colleagues and I have had several years experience using calculators in mathematics instruction at Ohio State. Calculators were first introduced in the remedial sequence in 1974. Now scientific calculators (with algebraic logic and hierarchy) are required in all courses below calculus, and one version of the calculus sequence uses computers and programmable calculators. The mathematics placement exam has been adjusted so that students may use calculators on it if they wish.

In this paper I want to discuss briefly the effect that using calculators has had on our teaching of mathematics to college-age students who have serious deficiencies in mathematics. We have taught many of these students as freshmen at the University, and in the last couple years we have worked with them as high school juniors and seniors. We have been limited in this course development by the goal of preparing students for standard precalculus courses. Our efforts have been less with defining alternative course content than with finding alternative ways of approaching traditional mathematics to make it accessible to larger numbers of students. It is the change of approach made possible by calculators that I want to describe in this paper. Hopefully, what we have learned about the potential for calculators in the teaching of algebra and geometry will be useful to those who are thinking about other mathematics courses at the beginning level.

What We Think We Know

In the academic year 1983-84, approximately 60,000 juniors in 614 Ohio high schools were tested through the Early Mathematics Testing Program. Almost all of these students said they planned to attend two-year or four-year colleges. Forty percent of them tested at the lowest level, "level 5" by our designation. Among level 5 students only about 5% can correctly interpret $\frac{x^6}{x^2}$ (although 40% can evaluate $\frac{10^6}{10^2}$); fewer than 5% can choose from a multiple choice list a solution to a system of linear equations with integer coefficients. Thus we say that level 5 students have no usable understanding of elementary algebra. However, 90% of these students have successfully completed Algebra I in their schools and more than 50% were enrolled in Algebra II when tested in the winter or spring of their junior year. This is the bad news.

The good news is we have evidence that a high percentage of these students can be successful with this mathematics when the approach to it is changed. The approach we have come to use is a highly numerical one with key concepts suggested in concrete problem situations. This approach is possible because of the availability of handheld calculators.

At Ohio State approximately 75% of the students who move from the remedial sequence into university level courses are successful in those courses. Although some of us might hope that this number could be higher, it is comparable to the success rates in the mathematics sequences at higher levels and is far better than when we taught arithmetic, elementary algebra, and intermediate algebra separately for this audience.

Of course, there are many reasons why students who are unsuccessful in high school may be successful in college. But we have demonstrated that students who test at level 5 as high school juniors can still learn much of the mathematics they need for college entry in their senior year. In 1981-82, we piloted an alternative course for high school seniors who had placed at the lowest levels on the junior year exam. The materials for this course were field tested in 1982-83 with more than 1000 students in 41 Ohio high schools; at the beginning of the year, about 750 of these students were at level 5, 140 at level 4, and 110 above level 4. (In Ohio level 3 is an acceptable university entry level for non-science students.) Over 80% of the level 5 students finished the course above level 5 with almost 70% above level 4; 91% of the level 4 students finished above level 4. Two-hundred of these students entered Ohio State in autumn 1983. Their mathematics grades during the academic year 1983-84 matched closely the grade distributions of the courses they were in.

The senior year high school course, like our university remedial sequence, takes a highly numerical approach to the traditional content of algebra and geometry. In developing these courses we have learned that a calculator can do much more than compute. It can provide access to key concepts of algebra and permit the numerical investigation of problem scenarios that otherwise would be inaccessible to beginning students.

We are convinced that there are better approaches with college-age students than starting with arithmetic algorithms and moving again through grades 5, 7, 9, and 11. This approach did not work for these students the first time. It suggests that students cannot do interesting mathematics until they can compute accurately with, for example, fractions. In algebra it requires students very early to use formal notation, to reason from axioms, and to handle greater abstraction than they are able to handle. It does not make use of their abilities to do mathematics within concrete problem situations. With calculators we can give up the fetish with paper and pencil computational skills and the formal, axiomatic approaches to elementary algebra. We can start with problems and calculators rather than with algorithms or axioms.

Using Calculators to Change the Approach

Our initial purpose in giving calculators to remedial students in 1974 was to get around arithmetic drill and practice, and to get students with weak computational skills into significant mathematics more quickly. There were other benefits. We immediately observed that calculators gave students more confidence and also enabled us to write more realistic and more demanding problems. These results were perhaps not surprising. Less predictable was the extent to which calculators provided an entrance to mathematics quite different from traditional approaches. Some features of this numerical, problem-solving approach are described below.

* Students with calculators can effectively use numerical methods (i.e. successive approximations or guess-and-check techniques) to solve problems that they are not able to solve algebraically.

With calculators, students are able to investigate problem situations in many special cases. Rather than merely attempting to answer a single question in isolation, students can investigate the relationships between quantities in a problem in many special cases, record information, and on the basis of this information answer many additional questions.

EXAMPLE: Water is added to 40 gallons of a 25% alcohol solution. Complete this chart to show how the concentration of the solution depends on the amount of water added.

Amount of water added (gal.)	Amount of new solution (gal.)	Amount of alcohol in new solution (gal.)	% alcohol in new solution
5	40 + 5 = 45	.25(40) = 10	$\frac{10}{45} \doteq 22.2\%$
10			
20			
			12.5%

> How many gallons of water must be added to 40 gallons of a 25% salt solution to give a 12.5% solution? How many gallons must be added to give a concentration less than 12.5%? How many gallons must be added to give a 10% solution?

Students become skilled in this type of problem solving and come to believe that with a calculator they can begin to solve most problems. Problems need not be limited to those that can be described by linear or quadratic equations. With calculators, beginning students can solve many problems they could not approach algebraically.

* Students using calculators can encounter basic concepts and relationships in problem situations before the concepts are formalized.

For example, in a problem about the growth of a bacterial culture where the doubling period is given as 10 hours, it is natural to ask for the number of bacteria after one hour. Initially, $2^{1/10}$ is a number the calculator gives a value for. Students can compute with fractional exponents before they deal conceptually with $a^{1/n}$.

The concept of function is a fundamental concept in algebra. If students only solve a single exercise in a given problem situation, they do not sense the underlying functional relationships in the problem. However, when they compute many cases, they do see the dependence of one quantity on another (i.e., in the first example that the concentration depends on the amount of water added to the solution). Calculators permit the extensive computation required to display these relationships.

Since algebra generalizes arithmetic, most algebraic concepts can be anticipated in numerical settings. This experience is particularly important for students who are not strong in mathematics.

* Investigating problems in many special cases permits variables to be introduced in a natural way to describe the general case.

The concept of variable is another absolutely fundamental concept of algebra. Our experience is that students who take algebra courses and two years later know no algebra are students who have little understanding of what a variable is or why it is introduced. If problem settings are investigated numerically in many cases, then introducing a variable to represent a set of numbers is meaningful.

EXAMPLE: Civil service employees anticipate a 7% raise. Complete the following chart.

Present salary ($)	12,000	15,500	22,000	25,400	X
New salary ($)					

Because students can solve problems numerically using calculators and guess-and-check procedures, the formalism of writing and solving equations can be delayed until extensive numerical experience supports these procedures. Thus, long before a student is expected to write and solve an equation in the problem setting above to answer the question, "If an employee's new salary is $24,610, what was her salary before the raise?" the student solves that problem experimentally:

Present salary ($)	12,000	15,500	22,000		
New salary ($)				24,610	34,240

* Calculators require students to pay attention to arithmetic ideas that otherwise would be of little interest to them.

Students must understand order of operations to use calculators. Division by 0 produces an ERROR display. Fractions are quotients. The subtraction key is different from the key that indicates a negative number. There is a square root key, a reciprocal key, a percent key. Scientific notation is necessary for displaying large numbers. Questions of how to recover a whole number remainder when one whole number is divided by another, how to use the calculator to raise a negative number to a power or to evaluate a polynomial at various values of the variable - these are natural questions that require understanding of arithmetic properties to answer.

In addition students are motivated to use their calculators efficiently, and they seek ways to shorten keying sequences. The distributive property is central in establishing algorithms for evaluating polynomials and also in simple computations: $12{,}000 + .07(12{,}000) = 12{,}000(1.07)$. Concepts of algebra such as factoring can be anticipated in efforts to make numerical computation efficient. Some arithmetic ideas that make sense with paper and pencil computations (like rationalizing denominators) are not needed and need not be taught when calculators are used.

* Calculators permit the early introduction of graphing so that graphing can provide a bridge between numerical work and algebraic representation.

In a numerical approach to algebra we choose problem contexts in which the relationship between quantities in the problem describes a function. These problem contexts can be investigated numerically through table-building and can also be displayed geometrically through graphs. It is important initially that students plot large numbers of points in drawing a graph. Level 5 math students usually do not view a graph as a collection of points, but rather as a few points that have been connected. Calculators make it possible for students to plot so many points that they realize if they were able to plot all, they would get the complete graph.

Graphing gives a visual representation of functions; students can handle graphs without the vocabulary of functions. Our students even graph calculator keys: [1/x], [$\sqrt{x}$], [x^2]. Again, important ideas can be experienced in concrete cases.

Graphing strengthens a student's sense of number magnitude and order. It is a natural setting for confronting problems of scaling and of ratio and proportion. This type of arithmetic experience, which can be artificial if done for its own sake, is a natural part of graphing.

Graphing provides an additional problem solving strategy for students. The same problem relationships that students deal with numerically can also be graphed. Once a graph is drawn, any number of questions can be answered -- including questions about inequalities. Students can solve equations and inequalities graphically before they learn techniques for solving them algebraically. They can, in fact, solve graphically (or numerically, for that matter) many problems for which they lack algebraic methods of solution. Calculators are important not only in drawing the graphs, but also in refining the estimates that are read from a graph.

Graphing can provide a bridge between numerical work and the formalism of algebra. Typically in these courses, problem situations are first investigated numerically, then revisited geometrically through graphs, and still later represented algebraically. For example, the problem about diluting an alcohol solution stated numerically on page 4 has geometric and algebraic formulations at later times:

Geometric: Water is added to 40 gallons of a 25% alcohol solution. Draw a graph that shows the relationship between the amount of water added and the concentration of the solution. (Use amounts of water from 0 to 80 gallons.) When is the concentration of alcohol equal to 12.5%? When is the concentration less than 10%?

Algebraic: Water is added to 40 gallons of a 25% alcohol solution. Let X denote the number of gallons of water added; write an algebraic expression for the concentration of the new solution. How many gallons of water should be added to get a 12.5% solution?

Deciding what a variable represents and using it to describe the relationship in a problem are sophisticated procedures and difficult for many students. The numerical approach seems to guide the student in understanding the relationships in a problem, and the graph provides a visual representation. Students see that the numerical table, the graph, and the equation are three ways of describing the same relationships within a problem.

Graphing can both suggest and support more advanced concepts of algebra and can be very helpful to students who do not handle abstraction well. Quadratic inequalities are questions about where a parabola is above or below the axis, completing the square is a process that yields the coordinates of the vertex of a parabola, the domain of a function consists of the numbers that give points on its graph, and so on.

A Next Step

A great deal of what we do in our courses for remedial students can well be considered pre-algebra. We are troubled that the transition from arithmetic to algebra is not being made successfully by large numbers of students in grades 8 and 9. In January, 1984, we began a project funded by SOHIO that will provide supplements for grades 7 and 8 in hopes of strengthening the pre-algebra component in these grades. The specific goal is to establish the concepts of function and variable before students move into algebra. Handheld calculators will play a central role.

Through several years of experience we believe we have come to understand ways in which calculators can give students entrance to many concepts of algebra and geometry. We have not had equivalent experience in using computers in the teaching of elementary mathematics. We hope to have an opportunity for some experimentation next year with students similar to those we have been teaching with calculators. Computers offer additional features, though it is not immediately clear how to exploit them in the teaching of algebra to a remedial audience. For example, we find that filling in several rows of a numerical table makes generalizing with a variable very natural for students. Can the same experience be provided with a computer program? Can, in fact, a student handle programming without already appreciating the role of a variable? If software is prepared can it be developed in a way that puts students fully in charge of problem solving? We have found that plotting many points to develop a graph strengthens students' number intuition and sense of number order. Will this happen if computer graphics replace calculator and graph paper? Students can begin using calculators with essentially no instruction. Will providing students with computers require us to sacrifice mathematics instruction time for programming instruction? Students have access to their calculators at all times. Will the more limited accessibility of computers make them less useful for instruction at beginning levels? These are interesting questions and there are many others.

Undoubtedly algebra must remain in the curriculum of two-year colleges. Courses in mathematics and in many other areas require algebra. For the foreseeable future we will continue to teach many recent high school graduates who do not know enough algebra to use it and also large numbers of adult students who have lost algebra skills they once may have had. We can use calculators (and maybe also computers) to approach this mathematics in a concrete way that is sensible to students and focuses on the solving of problems. I hope to see separate beginning college courses in arithmetic, elementary algebra, and intermediate algebra give way to courses that integrate arithmetic and algebra and make full use of geometry. The precise content of those courses need not be the same for all students. Two-year colleges are excellent laboratories for seeking new approaches that will enable more students to learn more mathematics.

Bibliography

1. Leitzel, Joan R., "Improving School-University Articulation in Ohio," _The Mathematics Teacher_, November, 1983, pp. 610-616.

2. Leitzel, Joan R. and Alan Osborne, "Mathematical Alternatives for the College-Intending," 1985 NCTM Yearbook, to appear.

3. Waits, Bert K. and Joan R. Leitzel, "Handheld Calculators in the Freshman Classroom," The American Mathematical Monthly, 83 (November, 1976), pp. 731-733.

(DISCUSSION BEGINS ON P. 288.)

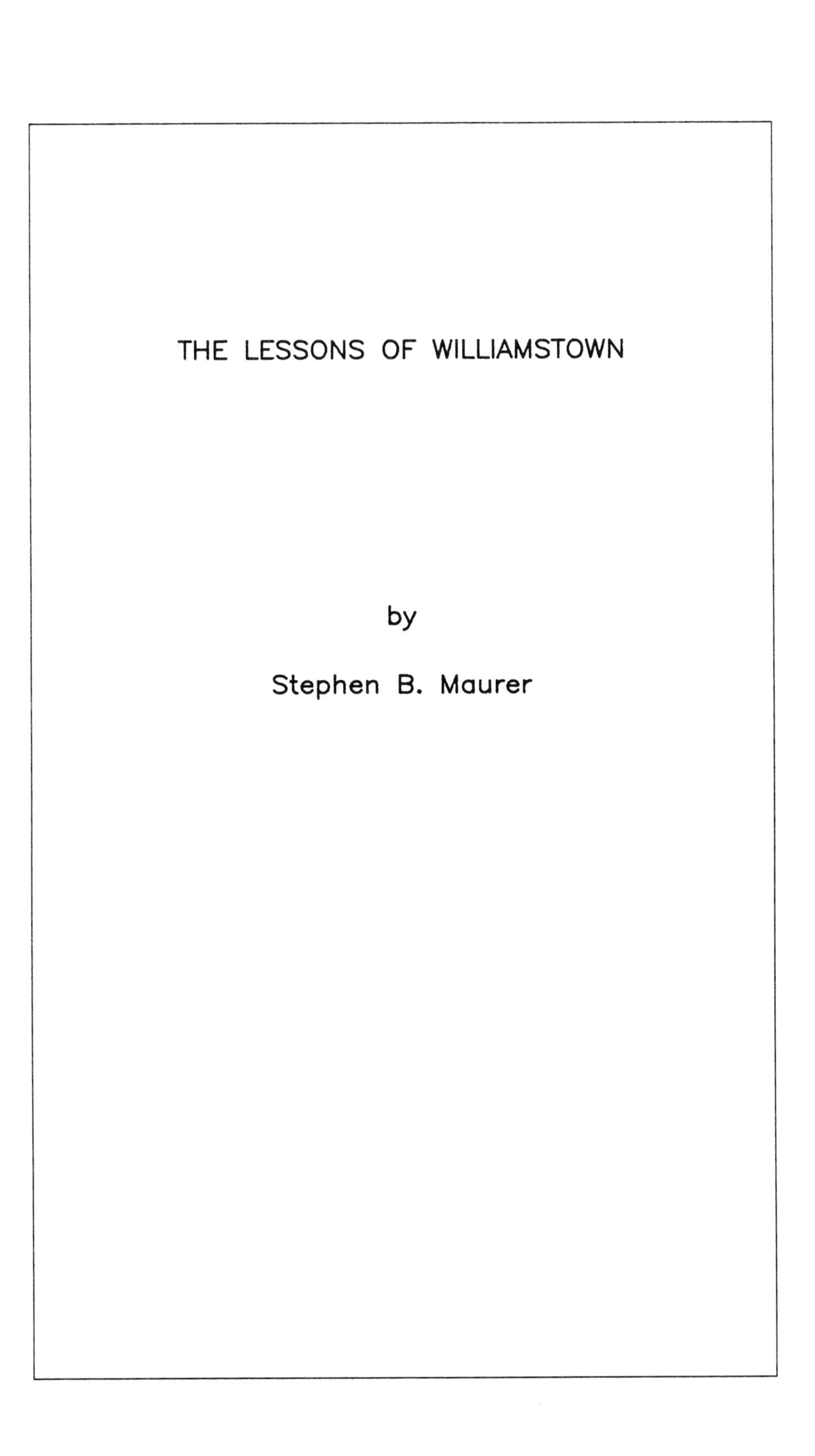

THE LESSONS OF WILLIAMSTOWN

by

Stephen B. Maurer

SUMMARY

The Williamstown conference of July, 1982, provided a detailed rationale and agenda for increasing the role of discrete mathematics in the first two years of the core undergraduate mathematics curriculum. What has happened since? In this paper, the author reviews what has happened, discusses some of the issues raised, and ends with some brief suggestions about the relevance to two-year colleges.

THE LESSONS OF WILLIAMSTOWN

Stephen B Maurer

The title was suggested by Don Albers. It refers to the conference held in July, 1982, at Williams College to discuss the need for more discrete mathematics in the first two years of the core undergraduate mathematics curriculum. This conference is a follow-up to that one, in the view of both the organizers and the Sloan Foundation, which has funded both. Consequently, it is appropriate to review the lessons of Williamstown as we begin in Menlo Park.

Unfortunately, the title has caused me a lot of agony. I'm not sure I have any lessons to tell! This should not be surprising. Given that a major overhaul of undergraduate mathematics was proposed, two years is a short time to see much happen. Perhaps I should have changed my title to "Status Report". Even then I'm not sure I have much useful to say. In this business, two years is a short time even to see any trends. But the fact is, part of the reason there hasn't been much change is that there is a lot of resistance -- not so much outright objection as severe doubt that the proposed changes can be effected successfully.

In any event, here's what I'm going to do. First I will say briefly what happened at Williamstown and where things stand now. Then, in a long middle section, I will give my analysis and reaction to the doubts. (Perhaps I should have titled this paper "Everything I am Thinking about the Discrete Math Project Now".) Finally, I will close with some brief thoughts about how all this pertains to two-year colleges.

Williamstown

The desire for change at Williamstown was unanimous. But then, this shouldn't be surprising. Had the title of the conference been "The Continuing Importance of Calculus", there still would have been a consensus for change -- nobody is ever satisfied with the calculus course!

More significant was the agreement that there should be more discrete math in the first two years of the college curriculum. But there was less agreement as to what discrete math to add and how. Should it be an emphasis on formal algorithms and recursion? Should it be math modeling? Are new courses necessary or is it enough to replace a few weeks in the current calculus and linear algebra sequence? All these were suggested.

There was also disagreement on what could be taken out. Could calculus be reduced to one year or not? Not surprisingly, the conferees from different fields (physics, engineering, computer science, social science) held different views. (Here and elsewhere, when I refer to a one-year calculus, I mean calculus including multivariate. That is, I refer to the material that usually forms a three-semester sequence or, in some places still, a four-semester sequence.)

Current Status

Interest in the mathematical community in this issue has grown by leaps and bounds. For instance, at MAA meetings talks and panels on discrete math in the first two years get SRO audiences. I sense that there is finally a pretty good understanding in the community of how the proposed algorithmic discrete math course differs from discrete structures and finite math. There is also a growing awareness that the issue is much broader than service to the computer science community.

There has also been some specific activity. Based on Williamstown, the Sloan Foundation invited several schools to make proposals for developing a new underclass curriculum, and six schools were funded at $40,000 each: Colby College, University of Delaware, University of Denver, Florida State University, Montclair State College, and St Olaf College.

Also, the MAA set up a Discrete Math Panel, headed by Prof Martha Siegel of Towson State (MD), to write a report much in the style of the Mathematical Sciences Report of a few years ago (part descriptive, part prescriptive). A preliminary report should be out this fall, a final report a year later.

But other than this, I think I have to say that not much is happening. Except for the Sloan schools, I haven't heard much about institutions revamping their curricula. Even the Sloan schools, which are in the middle of two and three year schedules, are having some trouble. They have no problem introducing a discrete course (just lots of work to produce text materials), but their commitment to reducing the calculus seems to be eroding. Even at the school which has placed the highest priority on how they redo the calculus (Colby, where the first year of the project was devoted solely to this, and a discrete course will be introduced next year for the first time) there is now the feeling that their one-year calculus course is really successful only for students who have had some calculus in high school.

I would like to point out an important exception to my statement that not much is happening elsewhere. Dartmouth has voted to change its program, effective for September's freshmen, to one year of calculus and one year of discrete math and linear algebra. In more detail, the core math program in the first two years for scientifically inclined students will be: a term of calculus followed by a term of discrete math followed by two terms which interweave calculus, including multivariate, and linear algebra. This sequence will be required for entrance to the math department. The first two terms of the sequence will be prerequisites for the _first_ computer science course. (This first course is quite sophisticated and assumes lots of programming experience. There are no programming courses per se at Dartmouth, but many courses, including the first calculus term and the discrete term of the new math sequence, include lots of computing.) In order to make this change politically acceptible to all departments, engineering and some science students will be allowed to take the second term of calculus without the discrete term, and there will be a special introduction to computer science for those with this background. However, the computer science department hopes this special course will be temporary (until engineers see the usefulness of discrete math). In any event, no student will be able to go any further in the mathematics or computer science curricula without that second term discrete math course.

Also, it turns out that at least two schools -- Grinnell and Williams -- are already doing calculus in one year. Indeed, they've been doing this for several years, quite aside from the discrete math issue. Both these institutions are now instituting semester discrete math courses to be taken after calculus, or for some students, after a semester of calculus.

So calculus certainly can be reduced, making room for discrete math, at least at certain sorts of schools. We have several constructive proofs!

Back to Dartmouth: there are actually two things going on. There is the new math curriculum for math and science majors, discussed above, and there is a new version of the finite math service course. Rather than writing a 4th edition of Kemeny Snell and Thompson [1], Kemeny Kurtz and Snell have written companion notes [2] introducing computer programming and providing lots of nontrivial mathematical examples to be carried out on the computer. (Write to Kemeny for a copy.) In other words, the course is algorithmic in the sense that students write and use algorithms, but they don't confront the mathematics _of_ algorithms, except informally. Although this is not the curricular change the Williamstown conference had in mind, it may well be a change which will succeed. (In order to make room for the new material, the course is also extended to two terms.)

Ruminations

Let me begin by returning to Williamstown for a moment. I may have given the false impression that it didn't accomplish very much. As you can imagine, during my two years at Sloan I have been invited to quite a few conferences on what to do about this or that. In retrospect, Williamstown was one of the very best. Most of the discussion was on specifics; there was little vague talk.

Moreover, specific proposals came out. Three curricula were proposed in the papers written before the conference, and two more were developed as consensus documents during the meeting. (One was for a unified discrete/continuous two-year sequence, the other for a curriculum of separate courses.) If you read these consensus documents, you will perceive much less disagreement than I intimate above. However, as I see it now, the reason agreement on what to put in was easy is that it was agreed to put almost every discrete suggestion in. The way it was agreed to take a lot of calculus out was to make the statement, almost offhand, that those who needed the things omitted could learn them later. This form of agreement may have been too facile.

Nonetheless, I do feel the contributions at Williams were high quality. To prepare myself to write this paper (or maybe it was to postpone writing it) I reread the Proceedings [3]. Yet again they provided me with new insights and fresh ideas. For instance, I was struck this time by how much the proposals involved integration of discrete and continuous ideas, e.g., teaching difference and differential equations concurrently. Yet, only one of the course development proposals solicited by Sloan proposed to try an integrated curriculum, and it was not well enough thought out to be funded. It seems that an integrated approach is the natural way one thinks to do things in the best of possible worlds, but that this approach is too great a leap when we actually commit ourselves to action.

As long as I am talking about new insights from rereading, let me mention another case. I recently had occasion to look carefully again at that classic hard calculus book of the 60's: Apostol [4]. My new insight was: it's not a calculus book. It has all sorts of things in it that we discrete math advocates have been talking about -- combinatorics, induction, probability and statistics, difference equations, numerical analysis. Granted, much of this is in the second volume, certain current perspectives and topics are completely lacking (the algorithmic perspective, the topic of graph theory), and it is all wrapped around calculus as the core. Nonetheless it's clear that Apostol wasn't trying to teach just calculus; he was trying to teach mathematics, as an integrated two-year course, according to the best judgment he could make at the time. This was very much the spirit of Williamstown.

Let me turn, finally, to what I perceive as the reasons for resistance to change.

Discrete Mathematics. The main problem here remains the lack of a book. Faculty like the idea that there might be a core of discrete math of value to almost everyone who will use math, and which is teachable at the freshman/sophomore level, but they'll believe it when they see it. The second problem is that, except for computer science, no serviced discipline is asking for discrete math. The people who are using discrete math for research in, say, management science or sociology are not yet asking that this be taught to their incoming students. All this is mirrored by the fact that there are plenty of books for freshmen and sophomores with titles like "Discrete Math for Computer Science", but none with "for Computer Science" deleted.

Actually, I did finally see the shorter title a few months ago, by Richard Johnsonbaugh of DePaul [5]. Johnsonbaugh makes the point in his preface that discrete math is good for a lot of people, though his topics still have a fairly exclusive CS bent.

In any event, a lot more books with<u>out</u> CS in the title are being written, by people at the Sloan funded schools, at Dartmouth, and elsewhere. If Tony Ralston and I would only stop going around giving talks, we might even finish our own book.

Calculus in One Year. This hasn't happened because most mathematicians simply don't believe it can be done. And frankly, those of us advocating it have been a bit naive -- at least I have. When people have asked me <u>how</u> it is to be done, I have made statements like, "You really could cut back on methods of integration or related rates". They counter by pointing out that they have <u>already</u> cut back substantially on these topics, and at most they would gain two or 3 days by cutting further. Furthermore, they point out that colleagues in the traditionally serviced departments are already unhappy that certain material their students used to know from calculus can no longer be assumed. When I say, "Couldn't you cover curve sketching by computer graphics rather than calculus?", they counter by arguing that it will take just as long for students to master the software and develop "function sense" this way as by the traditional method. When I say that symbolic computation software will drastically reduce the amount of time needed for drill, they are doubtful. Also, they don't have the software.

[A status report on such software: The fact that it exists does seem to have become well known. A "Computer Algebra" conference was held in April in New York at which some distinguished mathematicians lectured on how they have started using computer algebra in their research. The conference was expected to draw a few people: 400 came. As for the software itself, the number of powerful systems is growing -- MACSYMA, SMP, Maple, Cayley, ScratchPad, Reduce -- and the cost to academic institutions is down to a few K or less. (The basic cost of MACSYMA for a Vax is now $500.) But it's still true that these are big systems. They generally need a mainframe, on which they take a lot of space and/or run slowly if supporting several users. The only serious system for micros, MuMath, is much less powerful, and not very user friendly. The hand-held machine foreseen by Wilf [6] is nowhere in sight. In short, symbolic computation is not yet convenient for colleges. Also, textbooks geared to teaching with it simply don't exist.]

At any rate, the doubts expressed about doing calculus in one year have some merit. Such doubts have pretty well convinced me that 1) calculus cannot be shortened by nibbling away here and there, and 2) calculus is not likely to be substantially condensed by a productivity gain due to technology.

I still do think calculus can be reduced to one year in the first two, but one has to take a more radical approach. I will speak to this below, but first I want to deal with a different sort of objection.

The further objection is philosophical. I can best explain it by an example. In the last year, I have twice heard a distinguished analytic number theorist criticize Donald Knuth as follows. In "The Art of Computer

Programming, Vol 1" [7], page 58, Knuth gives 6 identities for sums of products of binomial coefficients. He relegates the proofs to the exercises; his hints suggest that he regards the proofs as varying from identity to identity, and being a bit tedious. The number theorist says that, if Knuth only knew the theory of hypergeometric series better, he would recognize that all 6 formulas fall out at once. The general point, of course, is that there is a long tradition in analysis, and one who doesn't know it all may have to reinvent the wheel, or may never discover that there is a wheel. If calculus education is diminished, all scientific progress will thus be slowed.

A related argument goes as follows. Haven't we advocates of discrete math already gotten what we need? Isn't discrete math already a popular offering at the upperclass level? The next generation of leading mathematicians are already being exposed to it.

I feel these arguments involve an important confusion. I think we all agree:

It would be best if everyone knew everything!

Unfortunately, that will never happen. Well then, can't we at least insist that future mathematicians should know a lot of analysis? By all means, we can, do, and should. But, at some point, there are diminshing returns. It's not clear that it would have been worth it for Knuth (who knows a great deal of analysis) to take an extra course in hypergeometric functions if all it saved him and his readers were a few pages of calculations.

As to the upper level discrete offerings, if all we were concerned about were the best students going into math, or even if we were only concerned about math majors, then it would be enough to have discrete math at the upper level. (It would also be enough to have continuous math at the upper level.) But there is a whole population to be educated; faculty at two-year colleges know this better than most. Moreover, students have the annoying habit of dropping out of mathematics at the most inopportune times -- sometimes after only one college course. Some never sign up at all.

In short, we educators are confronted with the following optimization problem. If 90% of students need to know something about A, 80% need to know something about B, 70% need to know about C, etc., and furthermore, most students take a limited amount of mathematics, in what order should topics be taught to maximize the amount of needed knowledge which students have? It seems pretty clear that a greedy algorithm is called for: to the extent that logicial sequencing permits, topics should be taught in the order of their breadth of usefulness. (Caution: I don't have a theorem here, because I haven't fully specified the model. For instance, it makes a difference which students drop out when.)

Let me return now to the matter of calculus in one year. There is a whole tradition of expectations about what students will learn in their first two years of mathematics. For instance, it is assumed they will learn integration by parts, but not that they will learn any probability. Consequently, a teacher in a later math course, or in another discipline, will do an integration by parts without starting from scratch, but must start from scratch with any probability. Similarly, a physics professor can assume that a student with 3 semesters of calculus knows Green's Theorem, but an economics professor cannot assume the student has even heard of Euler's Theorem on homogeneous functions.

If a major curriculum change is to take place, in particular, if calculus is to become one year, then there must be a conscious decision to make a wholesale change in such expectations. A lot of material must either be kicked up (it becomes grist for junior/senior math courses) or kicked out (it becomes an application which other departments must teach as needed). After some initial objections from colleagues who lose favored status, this all ought to be workable so long as everybody knows the score.

Here are my ideas about things to kick up or out. I've searched my soul, and without shame I can say to my colleagues, in mathematics and other departments, that:

1) Freshmen and Sophomores need only know about the simplest integration methods (simple substitutitons into the basic differentiation formulas, and maybe a

few examples of integration by parts; the rest should be kicked up). They need not be facile at doing any complicated integrations or differentiations by hand (kicked out).

2) These students need not know about convergence of series except the minimum amount on the ratio test needed for Taylor series (rest kicked up, or, if the first discrete math course includes generating functions, placed there in part).

3) These students need not know about partial fraction decomposition of rational functions (kicked up, or kicked into the generating functions part of discrete math).

4) They need not know about parametric representations (kicked up) or polar and spherical coordinates (the former kicked up, the latter out).

5) They need not know quadratic analytic geometry beyond circles and vertical parabolas (rest kicked out).

6) They need not know trig formulas beyond sin and cos of a+b (kicked out).

7) In multivariate calculus they should know about partial derivatives and about the classical total derivative, but not about the derivative as a linear transformation (kicked up).

8) Theys need not know about abstract linear algebra, but only about constructive linear algebra in R^n (rest kicked up). This, of course, is not part of cutting back calculus; it is part of cutting back linear algebra so that it fits in the new one-year discrete course.

Such kicking really will result in a shorter calculus course. I know this because the books are already around: they go under names like "Short Calculus". The only mismatch is that these books have been aimed at weak students, not average and strong students.

Such kicking means the upper class years for the major must also be changed. Students will not get as far in analysis as the ones we sent to graduate mathematics departments in the last 20 years, but they will be much broader. It is a trade I am quite willing to make, even for the ones who still do go on to graduate departments.

In part I am willing to make the trade because I have high hopes for the increased understanding students will obtain in the first two years, if we do the new curriculum right. In calculus, we must see to it that they have a good feel for the behavior of standard functions, and for interpretations of derivatives, differences, integrals and sums. In linear algebra, they should know a lot about setting up and solving linear equations. More generally, they should have a good grasp of mathematical language. They should also have a good grasp of algorithmic language and ideas, and know something about intelligent use of computers. They should have experience with a wide variety of optimization problems and techniques, some of them heuristic. Finally, they should have experience with discrete topics like graphs and logic, and discrete/continuous topics like probability and statistics. If some understanding of all this is gained, to the point of having a reasonable grasp of how to apply it, I think other departments too will feel the change in curriculum is a worthy trade.

The idea promoted above -- substantial excisions from calculus, not just streamlining, are necessary to make a one-year calculus course, and some of the excisions should reappear in the upperclass years -- is in the Williamstown papers. It's just not put in flashing lights. I think one lesson is that it needs to be.

Suppose radical change in calculus doesn't happen. Might there at least be more gradual change? For instance, might one incorporate in the calculus sequence the discrete approach to limits, the simultaneous treatment of difference and differential equations, and more numerical analysis algorithms? Yes, I think such changes will happen, whether or not more radical changes also catch on. Actually, over the years a number of calculus books with such ideas have been published. They haven't succeeded. But maybe the time wasn't right. Computing wasn't familiar and convenient enough, to the average faculty member as well as to students. Now it probably is. Incidentally, a recent effort along these lines, with reported good success, is the calculus course developed by Sheldon Gordon of Suffolk Community College, NY. He has written a text, but it's only available in note form. (Prof Gordon will attend this conference.)

Relevance to Two-Year Colleges

I have no expertise on this issue. Only recently have I come to understand (thanks to written and oral tutoring from Don Albers) that two-year and junior colleges vary considerably as to their educational objectives. I'm looking forward to gaining further understanding at this conference. In any event, I'm in no position to do more than make some general remarks on the relevance of the discrete/continuous debate to junior colleges -- remarks which may be obvious to you anyway.

Clearly, the relevance is different depending on whether a student is in a program aimed at transfer to a four-year institution or not. In the former case, you are pretty well constrained; you can't change your curriculum ahead of the four-year colleges and universities. Interestingly, in at least one state the constraint also goes the other way. Bettye Anne Case will report on how, in Florida, a university cannot require for admission to a department a course which is not offered at sufficiently many of the state two-year colleges which feed into that university. This means that either the two types of institutions work together on change, or it can't happen. (Bettye Anne is trying to ensure the former, which is certainly more fun for junior college faculty than waiting to follow the curriculum decisions of others.)

But what about nontransfer programs? Those of you involved in such programs have a wonderful opportunity. There may be even more incentive for you to increase discrete math than for the four-year institutions, and fewer roadblocks. I suspect that few of your students plan careers in engineering or hard science. Many of them may not need much calculus, certainly not much of the harder or deeper topics which would get left out in a one-year calculus course. So you can concentrate on revising the calculus and putting more discrete math in.

In his paper for Williamstown [8], Albers suggests that discrete math may have an additional value for remedial students, an especially significant group at many two-year colleges. Don argues that discrete math gives faculty something new to teach these students, accessible to them and interesting to learn, instead of teaching them the same things over again, only louder.

It seems to me we have to distinguish two extremes of students who place into remedial courses. First, there is the student who didn't learn the high school material for reasons independent of capability: the school was lousy, the home situation was chaotic, the student wasn't interested in learning then, the student had unfounded anxiety, etc. Second, there is the student whose ability is low: no matter how many times signed numbers are explained and practiced, no matter how much is done to overcome anxiety, he/she just can't keep the mechanics straight.

The first sort of student, assuming the motivation is now there, probably would prefer to see the old stuff again, and master it. Adding new things as well, which better illustrate the usefulness of math, is fine so long as it is clear to the student that the old stuff is there too. The second sort of student, however, is not going to gain from seeing the old stuff again. Of course, the problem is how to tell the two types apart at the time of course registration.

For the second type, I'm not sure discrete vs continuous is the right issue. The issue should be: is there some way to teach these students concepts while leapfrogging the computations which they cannot do? The obvious answer is computers -- symbolic computation, graphics, etc. It's certainly worth a try; not much else has worked.

But one must not be blindly optimistic. I pass along the observation my friend Joe Malkevitch (York College, CUNY) made when I suggested symbolic computation would save the day. Any software which does a subtle process must itself be somewhat subtle to use. For instance, differentiation being subtle, it's not surprising that (Df)g is different from D(fg), in which case the software must treat them differently and thus the person using the software must understand the difference. A person for whom the process is too hard may well find the task of correctly feeding the software no easier. If so, a software-aided course will get no closer to teaching concepts than the old course -- it will simply bog down in pushing buttons instead of pushing pencils.

Let's hope this pessimistic vision is wrong. In any event, there is an opportunity for two-year colleges to be in the forefront of finding out.

REFERENCES

1. John G Kemeny, J Laurie Snell and Gerald L Thompson, "Introduction to Finite Mathematics," 3rd ed, Prentice-Hall, Englewood Cliffs NJ, 1974.

2. John G Kemeny, Thomas E Kurtz and J Laurie Snell, "Computing for a First Course in Finite Mathematics" and "Computing for a Second Course in Finite Mathematics," mimeographed, Dartmouth College, Hanover NH, 1983,

3. "The Future of College Mathematics: Proceedings of a Conference/Workshop on the First Two Years of College Mathematics," Anthony Ralston and Gail S Young, eds, Springer-Verlag, New York, 1983.

4. Tom M Apostol, "Calculus," Vol 1, 2nd ed, 1967 and Vol 2, 2nd ed, 1969, Wiley, New York. (Original editions published by Blaisdell, 1961 and 1962.)

5. Richard Johnsonbaugh, "Discrete Mathematics," Macmillan, New York, 1984.

6. Herbert S Wilf, Symbolic Manipulation and Algorithms in the Curriculum of the First Two Years, in [3], pp 27-42.

7. Donald E Knuth, "The Art of Computer Programming," Vol 1, 2nd Ed, Addison Wesley, 1973, pp 58.

8. Donald J Albers, The Impact of a New Curriculum on Remedial Mathematics, in [3], pp 191-199.

(DISCUSSION BEGINS ON P. 299.)

STYLE VERSUS CONTENT: FORCES SHAPING THE EVOLUTION OF TEXTBOOKS

by

Peter Renz

SUMMARY

To focus on matters of content to the exclusion of matters of style and execution in proposing new curricula is to lose sigh of the essentially evolutionary nature of the development of courses and course materials. Strong selective pressures work against large-scale changes in large courses. These constrain are partly the result of inertia and of existing requirements outside of the mathematics curriculum, and they are partly economic in nature, affecting what will and what will not be economic to publish. This is a review of these limitations on curricular reform and the conclusion is that the success of such changes is more dependent on the style and execution of the materials produced than on the exact content of those materials.

STYLE VERSUS CONTENT: Forces shaping the evolution of textbooks

Peter L. Renz
Division of Science, Bard College
and W.H. Freeman and Company

STYLE AND CONTENT

What is it that makes a truly successful course or text? Is it more a matter of content or of the manner in which the material is presented? The answer depends upon one's objectives and one's point of view. On balance I say that style and manner of presentation are more important than content. In particular, the matter of discrete versus continuous mathematics seems a side issue. The main issue should be how to achieve success in the classroom, and this issue depends upon the details of the teacher's interaction with his or her student. Where texts and courseware are concerned, success depends upon the ability of authors to reach out to students with apt and compelling arguments and with clear and evocative images. Success in education also depends upon students and teachers having clearly formulated goals, both overall and in detail, and ways to judge the level of success in achieving these goals.

One might properly be alarmed or horrified by the content, or even the methods, of R.L. Moore's topology classes. One might even disapprove of the narrow focus that comes with the use of pure discovery-learning, but Moore's successes and those of his students are legendary. This, therefore, shows that style can triumph over content. I venture that there are no examples of the triumph of content over style. If the presentation is sufficiently bad, the students are lost.

The debate concerning whether discrete or continuous mathematics should be central to the curriculum will simply pass away. Change will come by evolutionary forces. Those who heralded the discrete revolution will be enshrined as saints by some and cursed as devils by others. But the changes that are made will be the work of the foot soldiers, teachers in the trenches who are subject to unpleasantness and risk as they work up new course material or help purge the errors and infelicities of other's course notes. These teachers are my heroes, along with their striving and suffering students.

The discrete revolution is being oversold. I raise three objections to it as a cure-all. First, the discrete/continuous dichotomy is not as sharp as it is pictured. This is borne out by many sources, but I have been particularly impressed by the depth and breadth of the arguments made in response to Anthony Ralston's position piece to appear in the November 1984 issue of *The College Mathematics Journal*. The responses by James P. Crawford, Daniel J. Kleitman, Peter D. L Saunders MacLane, Daniel H. Wagner, and R.L. Woodriff emphasiz the central importance of the insights gained from continuous mathematics. Insights of great use *even for the study of discrete systems*. Such respondents as William F. Lucas and R.W. Hamming stressed the importance for modern applications of both discrete and continuous mathematics--and these respondents are very strong proponents of discrete mathematics.

Second, the importance of manner of presentation of the materi has been largely pushed aside in the struggle over what is to be presented. My thoughts on this have been sharpened by seei how clearly this issue has been set forth by the responses to Fred Roberts's position piece in the cited issue of *The Colleg Mathematics Journal* on the role of discrete mathematics in the college curriculum. Of the six responses that I have seen in draft, four see the main issues in the introductory college mathematics curriculum as being pedogogical (style and manner of presentation) rather than content. I direct your attention to the responses by John Mason, Patrick W. Thom son, and William Ellis, Jr.

Third, until strong and successful models for these new discrete mathematics courses are available, we do not even know exactly what is being proposed. The reason is that discrete mathematics is a very broad area and it is full of very difficult and demanding material (try to master Ramsey theory, for example). Until a practical course has been plotted through these seas, the proposal that this voyage be undertaken by large numbers of teachers is as irresponsible as it would have been to propose a general assault on the problem of sailing west from Europe to India before Columbus, Magellan, and others had led the way.

For these reasons and for others set out below, I believe that we should put the noise of debate behind us and get on to the real business at hand: making the experiments that will give us new courses for this new curriculum and make safe the voyage to this new land.

My plea is for evolution not revolution for two reasons: first, evolution is the way things actually happen and second, evolution is a continuing process. Change is essential if mathematics is to be a vital subject.

DEVELOPING NEW COURSES

The development of new courses from their conception is similar in some respects to the emergence of a new species. Ideas serve as a modifiable genetic code guiding the development of the course, but these ideas are not enough. There must be the proper local environment to allow realization of the idea; and, if anything is to come of this all, the idea and its realization must be able to catch on elsewhere. The proposal to restructure the college curriculum so that discrete mathematics is taught early and calculus comes later is like proposing to insert a new gene into the chromosomes controlling the cells of the curricular organism. I believe that a cautious approach should be taken to such experiments lest one produce monstrosities.

Natural mutations yield many variant forms, but those that

represent substantial variations of large creatures naturally abort and never live to see the light of day. Successful variants begin small. Mammals first appeared as a few small creatures and through evolution, with its general tendency toward larger forms (Cope's rule) and greater diversity and specialization, gave us the full range of mammalian life that we see today. For creatures the size of elephants, we see little in the way of rapid change. (See McMahon and Bonner (1983) for the biology.) So it should be with curricular change. The "new mathematics" experiment illustrates the disastrous potential of large scale, rapid, and radical change imposed from without.

The wonderous diversity of life arises from selective pressures acting on random mutations. We may best understand the dominant types of texts and courses as rising from a similar evolution under competitive pressures. This viewpoint explains much, even though it ignores the distinction between the wholly random character of mutation in biology (prior to the invention of genetic engineering) and the purposive nature of curricular reform, and course and textbook planning.

As an editor, it has been my task to anticipate curricular developments and to select or develop texts that will establish substantial positions in both existing and new markets. From the editor's point of view, it is clear that there are more sound, even exciting possibilities than there can be economic niches in the market. These possibilities, whether put forth in book proposals, class notes, comprehensive plans for curricular reform, or in discussions, are just so many hopeful mutations whose potential is yet to be proved. This presumes that we know what success means in this context. For publishers, success is a commercial matter—sales great enough to cover costs and give a pleasing return that can be used to finance new projects, provide increased compensation for employees and increased returns for investors, and meet all the other needs of a prosperous enterprise. What publishers seek is overall success-that is, success of the enterprise as a whole. This is by no means inconsistent with taking on high-risk projects of occasional projects primarily for the good of the discipline and as a sign of the house's commitment in a given area.

Academics and publishers often do not understand each other simply because they do not understand each other's goals or their separate conditions of life, including the constraints on achieving their separate goals. Here I will give the picture as seen by a publisher. I will focus on the numbers, because they are easy to understand and they are the main constraints affecting all publishers almost equally. Moreover, the numbers will show where the best opportunities for curricular reform lie, insofar as realizing those opportunities depend on the publication of new course materials

WHY LARGE TEXTS FOR LARGE MARKETS SHOW LITTLE VARIABILITY: THE EFFECTS OF SCALE

Large technical texts with many diagrams are expensive projects even before the first copy is printed. Typesetting, art, page make-up, including allowances for all proof and corrections for a two color, 8 X 10 inch, standard calculus book of about 1200 pages would cost about $228,000 in 1984. To these costs we add editorial expenses including reviewing, checking of answers and solutions, travel., grants to the author for manuscript preparation, and in-house editing and development. I estimate these at about $50,000. Finally, for a complete package one must add the cost of supplements (Solutions Manual, Student Study Guide, and so on), a small publishing program itself. These might run to $40,000.

How does such a book look as a business proposition? The publisher must face the fact that competition, including the new and used book market, takes a very heavy toll on sales in the larger markets (annual decreases in sales usually exceed 20% and may exceed 30%). Furthermore, substantial sales for mainline books can seldom be sustained much beyond the fifth year after publication. Thus, a book that sells 10,000 copies in this sort of market in its best year might be expected to sell 30,000 to 40,000 copies altogether before it fades completely. Such sales would prove disastrous for the publisher. Here are the numbers for sales of 35,000 copies over 5 years at a nominal list price of $40.

<u>Profit (loss) on sale of 35,000 copies of a calculus book</u>

List price ....$40.00, Net Price.......	32.00
Net receipts	$1,120,000
Less printing, paper, and binding	
costs at $5.11 to $6.50 per book...............	$178,850
Less royalties at 16.5% of net	$184,800
Less operating overhead	$560,000
Less investment	$318,000
Net loss......................	$121,650

Here I have allowed 50% of net receipts for all operating costs under the assumption that all miscellaneous costs, including those representing the time value of money are expensed for the publishing house as a whole. This allowance is perhaps high, but it is close to actual costs, and it is not clear that this allowance could be reduced below 40% for the operations of any sizable publishing company. I have assumed that inflationary and interest effects generally cancel each other. The exact cost of the printing paper and binding depends on the lengths of the printing runs. With a run of 30,000 copies the cost would be about $5.11 per copy; with a run of 10,000 or under copy cost could rise above $6.50.

For the publishing house to earn a profit of 10% of net before taxes (a modest rate) with this model, it needs to sell about 60,000 copies. Again allowing 5 years for this would require selling 17,000 to 20,000 copies in the best year. This represents very substantial success. In the real world few books sell substantially above 20,000 copies in their best years. The publishing house does not move into a really comfortable position until sales of 30,000 to 50,000 copies are achieved in a calculus text's best years. However, success in the 30,000 to 50,000 copies per year range allows important economies of scale. In particular, it floats a large operation (promotion, sales, etc.) and it positions a company well for revisions, which are generally less expensive (art can be reused, etc.) and more certain in pay-off.

Nevertheless, this analysis shows why in freshman calculus (500,000+ students per year) and other large markets competition produces uniformity. No publisher can afford to produce a full-scale calculus book that is so different or so specialized that it is not a serious contender for mainline courses.

In recent times, no slim alternative book has swept the market or even taken as much as 5% of the market. But the costs of launching a full scale calculus book can't be comfortably borne by sales of less than 5% to 10% of the market. While the upper end of this range, 10%, represents about the largest slice of the pie that one might look for today. The result is that calculus books tend to compete mostly through minor variations and improvements. The dominant books expand by spawning minor variants that act to cut down the profitability of the market and thus squeeze out competitors. Consider Swokowski's Alternate Edition.

Here we see Hotelling's law of duopoly at work. Large competitors compete by matching each other's products and going for the center of the market. Consider Ford and General Motors, Time and Newsweek. This is convergent evolution at work in an area where the economies of scale allow only a few variants to survive. The freshman calculus text is an elephant in the world of texts and the economy simply can't support very many types of elephants. Economic forces prevent substantial innovations in courses of this size.

CURRICULAR FORCES LEADING TO UNIFORMITY IN LARGE COURSES

Large courses are taught by teams and the direction of such courses is determined by group choices. Committees work by averaging and tend to conservative decisions. Moreover, large courses in the sciences and mathematics usually have a strong service function, both for higher-level courses in the same discipline and for courses in other disciplines. The client departments for such students expect a uniform and dependable product, and these expectations strongly limit the possibilities for innovation or change.

Redirection of the introductory curriculum requires agreement on new standards. Until such standards prove themselves workable in the classroom and are found more effective for student's future work and for the client departments, prudent departments will not rush to institute new courses and prudent publishers will not rush to publish large or lavish books for such courses. Indeed, even mild deviations that are untested may be viewed with suspicion.

COMMERCIALLY ATTRACTIVE INNOVATIONS

Today's calculus book evolved from smaller ancestors by gradual additions and adaptations. If the present type did not exist for us to imitate, no author could or would invent it as an entirity and no publisher would invest in it. However, new types of texts are produced every year. Many are eliminated by market forces. The quite considerable costs and development and publication are borne by the authors, their institutions, and the publisher. As an editor, I have seen that many serious efforts are never pushed to completion by the authors while others are found by reviewers to be unsuitable for their intended markets and do not see the light of day. Each of the following examples was innovative and commercially attractive. They run from small projects to immense ones, and they illustrate what qualities innovative projects must have if they are to be commercially acceptable.

Smaller projects allow more attractive opportunities for innovative publishing. For example, Loren Larson's ALGEBRA AND TRIGONOMETRY REFRESHER FOR CALCULUS STUDENTS was based on a successful supplement in use at the author's school. Neither the idea of a review book nor the concepts and skills to be reviewed (content) were novel, but the way in which the review was set out was compelling and the care in bringing it together let one see that this particular approach would work (style). This project was attractive in its existing form to reviewers at other schools. Furthermore, it made sense to print this informal book from author-supplied copy and typed in final form after the publisher's work on the art, design, and editings. The result was a project requiring an investment equal to about 0.05 times that of the standard calculus book described earlier.

To break even with this sort of small project, one need only sell 3,000 copies. The sales of this book have exceeded 30,000 copies in five years and it has achived a percentage return on net beyond that which could ever be expected in a standard calculus book.

Other examples of similar success of small projects based on successful texts in use at a local school, each of which meet a need not met by existing texts, are QUICK CALCULUS by Daniel Kleppner and Norman Ramsey (John Wiley and Sons) and OPERATIONS RESEARCH FOR IMMEDIATE APPLICATION: A QUICK AND DIRTY GUIDE by Robert E. Woolsey and Huntington S. Swanson (Harper and Row). I would argue that the success of these books is more a function of their style and execution than of their content.

An innovative project of larger scale is David Moore's STATISTICS: CONCEPTS AND CONTROVERSIES (W. H. Freeman and Company), again it was based on a successful course at the author's school and had proven attractive in manuscript form to those experimenting with new types of statistics courses elsewhere. This project represents an investment equal to about 0.10 times that of our standard calculus book and requires sales of roughly 10,000 copies to break even. To achieve a 10% return on net with this sort of project sales of 20,000 copies are needed (roughly 0.33 times those required for our standard calculus text). The slow pay-off here is a result of holding the price down to allow supplemental use, a strategy that succeeded very well. Larger still and equally successful in its niche is the innovative statistics book by David Freedman, Robert Pisani, and Roger Purves (W. W. Norton). This text was developed and class tested over many years at the University of California at Berkeley. It is style and execution not content alone that has made these books successful.

More pertinent to the discrete mathematics debate is MATHEMATICAL STRUCTURES FOR COMPUTER SCIENCE by Judith Gersting (W.H. Freeman and Company). Again, this is a text that arose from an existing successful course at the author's school. Here the target was the ACM's proposed course on discrete mathematics. The books cited earlier in this section were innovative in their goals; this is not the case for Gersting. This cours existed elsewhere, enrollments were rising, and the author was unsatisfied with the available texts. It was not a matter of what the available texts set out to do (content) but rather how they did it (style). Gersting's course notes were polishe during more than three years of class testing before the book was published. Her class notes were extensively reviewed by other teachers to ensure that the local success could be repeated at other schools. The result is a text that requires an investment not much above 0.10 times that of our standard calculus book and that shows a profit of more than 10% of net receipts with sales of less than 10,000 copies. That is about 0.16 times the sales required for equal profit from a calculus book. Moreover, sales of existing books indicated that conservatively one could expect to sell 5,000 copies per year if successful. Actual sales have been about twice this high. This project was pleasing in prospect and it has been rewardin in retrospect. This book's revisions will face a more competitive market and give both author and publisher smaller slice of a much larger pie.

W. H. Freeman and Company has published several books that set new patterns for what later came to be known as liberal arts mathematics: books by Sherman Stein (1963), Harold Jacobs (197 and Bonnie Averbach and Orin Chein (1980).

In each of these books, matters of content and style are inextricably bound together and each would have vanished without trace had it not been for the author's and publisher's superb execution. Each of these books was based on a successful working course and each stood the test of critical reviews before publication, two criteria that every innovative text should meet. Each has proved successful in the classrooms of many other teachers after publication. Each offered a fresh point of view and a clear alternative for adopters. Each seemed a sensible commercial gamble, although declining enrollments in liberal arts mathematics courses in recent years together with a proliferation of available texts have made this a difficult market in the 1980's.

Each of these books requires a prepress investment roughly 0.25 times that required for our standard calculus book. For the earlier books, those by Stein and Jacobs , I estimate that that investment would be fully recovered with the sale of 20,000 to 30,000 copies. For the most recent book, that by Averbach and Chein, the break even point for sales lies above 37,500 copies and had not yet been reached in 1984. Compared with a calculus book the investment is about a quarter and the sales requirement to break even is about half. This is an interesting sort of gamble. The profit as a portion of net receipts lies in the range of 10 to 15% for the fully mature older books. This relatively low rate of return comes from competitive forces. Long-term survival drives companies to offer the best product they can at a price low enough to keep the competition down. The spread between the marginal cost of printing, paper, and binding and gross receipts is smaller for books that have many competitors than it is for those that have few. This is one reason why more advanced books are more expensive; the relatively smaller size of the markets for advanced books is another.

Large-scale innovative textbook publishing is so risky that it becomes attractive only when much of the costs are carried by others. The committment of the professional community and of the government to The New Mathematics and to Chem Study carried forward these two innovations. I cannot cite any figures for The New Mathematics, but I can cite some for the initial Chem Study materials published by W. H. Freeman and Company in 1965. The rate of return on net sales has been very modest but the net receipts over nineteen years have exceeded eight million dollars and the investment was low. This chemistry reform project was meticulously developed and backed by the full faith and credit of the community of chemists and the government. It was a success in the classroom and spawned a new generation of successful texts.

The terms of the arrangement allowing W. H. Freeman and Company to bring out this material forbade the Company from revising the original Chem Study material and hence from using its position to gain an unfair advantage once the new curriculum was established. It seems likely that the greatest rewards in this area were reaped by other publishers who profited by following the path broken by Chem Study and made improvements on the prototype published by Freeman.

NEW CURRICULAR MATERIAL AT THE DAWN OF THE AGE OF COMPUTERS

While the debate on continuous versus discrete mathematics continues, computer technology is creating a new environment. Students flock to existing discrete mathematics courses at the sophomore/junior level. These courses are part of the recommended curriculum of the ACM and they meet, in part, real needs for computer science departments. How some of this material can be worked into other courses remains to be seen. I would hope that the portions of it that are calculus related (generating functions, expectations, formal expansions, analysi of certain algorithms, etc.) will show up with more emphasis in introductory calculus.

Computer technology in the form of TEX, TROF, EQN, and other typesetting packages, combined with powerful word processing programs and graphics packages would enable departments to develop far more polished preliminary editions of experimental texts. This may allow new books to be produced at substantial savings. The result is likely to be an unprecedented flourishing of innovative and experimental course materials. The disadvantage will be that this will increase the amount of course material to be sorted out by the marketplace. Normally, publisher plays an important role in sorting out projects and helping to improve them. As the means of production move more into the author's hands, this role will be reduced.

The main issue is not whether or not there will be more discrete mathematics in the curriculum, but how the detailed matters will be settled: What discrete-related material will come to receive more emphasis in existing courses, how new discrete courses will come to be organized, and over all, how compelling a mathematics curriculum can be devised and put into effect. The realization of any change is bound by economic constraints, some of which are outlined here. From my view it is style and execution that are of central importance for, as an editor, my job is to find the best author and to help that author produce the best course materials. In this context the content is almost a given and the differences in execution ma the difference between success and disaster. Thus the details of style and execution are central for me, and I represent the editorial decision point. Whatever program is devised must at some point pass editorial judgement of one sort or another. Here I give you an outline of the financial constraints on such judgements.

The financial constraints are simply the conditions of life for publishers. How do publishers determine what projects are likely to meet their needs? I have given some of the criteria, but I have left out others that are obvious and extremely important. Is the project exciting? Is the conception and organization compelling? Is it clear to those within the company and to reviewers that the author has a real contribution to make? Can the virtues of the product be made evident to your intended audience? The last is essential to achiev reasonable sales. We are dealing with course materials, and so it is natural for potential publishers and for potential adopters to ask for proof of success in the classroom. The questions in this area are: Has the material been class tested? Does it work? Does it appeal to other teachers? Finally we come to the economic questions: Is there an existing or prospective market large enough to make the project economically feasible: True, content is part of all of this evaluation, but success or failure of a book is not usually a matter of content--content is too obvious a matter. Success comes to authors with a gift for exposition, who can give the attention to detail that makes a book work for the author, the author's students, and for others.

The computer age will bring us new ways to produce books and new ways to put together instructional materials (courseware). These developments bring the author closer to the role of compositor and allow the author to move into a realm previously the province of the publisher. This has the potential of greatly reducing the publisher's prepublication investment in new teaching material and thus allowing for less costly innovation. This means that the real costs of creating and sorting out new materials will fall more on the authors (who will collectively produce greater quantities of new materials for limited markets and as a result must, overall, earn a lower averag return) and on the users (who will be confronted by a larger selection of less-carefully tested products). These developments seem inescapable consequences of present trends.

As these new courses, texts, and other materials are developed, the content will largely be determined by the perception of student needs, but successful courses will always depend critically on the style and presentation of the teacher in the classroom and of the authors of the materials used. Successful teaching is done in detail, not by choice of content and large-scale strategy.

REFERENCES

Crawford, James P. 1984. Calculus is not an indescretion. _The College Mathematics Journal_, November 1984.

Hamming, R. W. 1984. Calculus and discrete mathematics. _The College Mathematics Journal_, November 1984.

Kleitman, Daniel J. 1984. Response to Anthony Ralston's position. _The College Mathematics Journal_, November 1984.

Lax, Peter D. 1984. In praise of calculus. _The College Mathematics Journal_, November 1984.

Lucas, William F. 1984. Discrete mathematics courses have constituents besides computer scientists. _The College Mathematics Journal_, November 1984.

McMahon, Thomas A., and Bonner, John T. 1983. _On Size and Life_. Scientific American Books.

Wagner, Daniel H. 1984. Calculus versus discrete Mathematics in OR applications. _The College Mathematics Journal_, November 1984.

(DISCUSSION BEGINS ON P. 306.)

DISCUSSION

NEW CURRICULA AND NEW TOOLS

1. Computers and Computing

2. Calculators

3. Discrete Mathematics

4. Evolution or Revolution ?

NEW CURRICULUM ELEMENTS AND NEW TOOLS

Fusaro: Now there is one thing missing in my tetrahedron model and I don't know how to get it in there. It's what I call "diagrammatic mathematics." That's a favorite term of mine; it's going to go by the wayside pretty soon, I think, and be replaced by something else.

Let me read from the table of contents of a book that was published in 1951 by Rule and Watts: "The straight line and the circle; the construction of conic sections; projected constructions; roulettes; vector geometry; graphical scales; empirical curves; periodic curves, including Fourier series and harmonic analysis; graphical calculus; geometry of projection drawing, axonometry"--I'm giving it away, I guess. Anyone want to guess what the title of that book is? That was eleven chapters, number twelve is "Conventions - Practical Drafting." That, dear audience, is a book on engineering graphics. In 1951 it couldn't be anything else, because this couldn't be considered part of mathematics. My view is that's as much mathematics as anything else. I think it's absolutely fascinating that engineering graphics at that time, in fact, involved mathematics. If you were to sit down with Mac Draw right now on a Macintosh, you could probably accomplish most, if not more, than the students who came out of that course could do. The mathematics hasn't gone away, but it's imbedded in the machine. It's down in there in an opaque way and you don't have to go through the process of learning all of that; it's in the machine now in the same way that there's some mathematics embedded in the calculator. Certain calculations you can do now by pushing a button--you don't have to carry it through in your head or push symbols on paper. You've moved up. As I mentioned earlier, you've elevated the level of cognitive activity and the mathematics has been pushed down out of sight due to a machine. I think that is an excellent metaphor for what's going on now and is going to continue to go on in the future, and I regard that as a fundamental fact that the curriculum has to come to terms with.

Gordon: If we are going to change the mathematics curriculum, how do we change it? It's got to be either very incremental with relatively minor changes that we can absorb that will not affect transfer options, or it's got to be wholesale. And I mean that literally; it's got to be a global change because it's got to be accepted on a nationwide scale at all institutions, by all Boards of Regents, and so on, because the changes are the full spectrum from arithmetic on up. That is something that I don't foresee happening imminently.

One of the great things about the community college faculty is that innovation in curriculum matters is the tremendous outlet for them. They're not involved in research activities and yet many of them hunger personally to do something. Ten years or so ago most of the development in personalized instruction was done at the two-year colleges. Within the last decade I think far more in the way of computer-assisted instruction of all types has come out of the community colleges.

But there are other problems. There is a confidence problem that exists with a lot of the faculty of these schools. Somehow the universities are an intimidation factor. We tend to look around to see what they're saying to us, or what they are doing, and we try to follow in their footsteps. Maybe a little more articulation with them might give us the self-confidence to more actively pursue some of these changes and see where they lead.

COMPUTERS AND COMPUTING

Tucker: The issue Ben Fusaro brought up before is one that may make inevitable what is otherwise impossible to move, that is, the computer. Students like to work with computers and this has proven to be a very useful tool in forcing people to do algebra--you can't program without algebra--the computer has this incredible power to make students keep on working until they get it right. It's a major step, and this brings in the issue of the discipline that is one of the underlying sociological underpinnings that is missing in many students' approaches to studying mathematics. The computer has important indirect aspects: Look at what happens when you try and get a program to run. Programs often don't run when you first write them. If you're doing a program that does anything mathematical and you get funny answers out, you compare what it is giving you with what it should give you. You have to do that by pencil and paper, and you end up thinking about the mathematical problem intensely to figure out where your program went wrong. So the underlying symbolic skills that we say are what's really happening in elementary algebra are going to happen if you teach a fair amount of numerical basic programming. Numerical methods bring in matrix algebra. There are all sorts of ways in which the computer is going to develop these skills and it's just inevitable that some of this is going to happen.

Smith: You seem to be equating the computer with programming. I am using the computer more and more but I'm doing less and less programming. The uses I make of the computer are with packaged programs. It's a lot easier for me to buy a program for $30 than it is for me to fool around trying to program, even if it's something fairly simple to do and something I know I could do. It's not worth it! We are going to see more and more pre-packaged programs. Now you can talk about the people that have to sell programs and certainly that's a marketing thing, but I think most of our students are going to make more and more use of packaged programs rather than writing the programs themselves.

Tucker: I agree, but there are lots of levels. Consider calculus--when you use calculus formulas, there's a huge amount of algebra that you get just for free when you plug in the derivative formula. We now teach elementary, first-year algebra to develop the skills students need later. Perhaps you teach beginning programming to develop the skills students need later, and then later on they work at them at a higher level.

Renz: In order for the computer to have these effects, the machines have to be essentially universally available, and I think that Allyn Washington in his view of the calculator and technical mathematics will bear out the fact that it was when the price of calculators went down, so that no one could afford to be without one, that the calculator became a universal adjunct to technical mathematics both for people working and for those learning about it. I don't really think that we're going to see this until you have essentially among serious college technical students universal personal ownership of computers.

Long: Don, you've been asking for recommendations. I wrote one out just before you said this, Peter, and it says this: "In view of the steady increasing availability of microcomputers and the urgent need to incorporate the use of micros in the teaching of mathematics, it is imperative that each two-year college mathematics faulty member be provided with a microcomputer in his or her own office. One of the primary deterrents to the use of computers in the teaching of mathematics is the necessity of most faculty to compete for limited facilities. This limitation must be removed. It is also necessary that these facilities be increasingly available to the students."

What I am saying in this recommendation goes further. I think the community college has the responsibility. It ought not to be just the largesse of industry. I think the community college has the responsibility of coming up with the bucks to provide the necessary equipment just like they buy microscopes and just like they buy other equipment. No longer is it the case that mathematics is simply pencil and paper, and we have a right to equipment just as much as the physics department or the chemistry department.

Gordon: I'd like to bring up a longer-term issue. Right now students in the elementary schools are being exposed to computers and to programming. If you picture ten years, say, on the average of development in these kids until they walk into our classes, they are going to be incredibly sophisticated. We have to start anticipating what we are going to face, and what we will have to offer them based on the knowledge they have.

Cohen: According to the latest figures, 43 percent of entering college freshmen have written a computer program.

Gordon: What percentage of college faculty have written a program?

Warren: My electronics faculty would not exist; they couldn't continue in their teaching positions if they didn't have it [programming experience]; and they got it on their own. I think that there's either a reluctance or a fear of increasing the rate of learning in some academic disciplines because people don't want to move ahead. It's not just math; it's other areas too. There's a fear of the content. If you move all that microcomputing and all that calculating back to the seventh and eight grade or the third and fourth grade, what are you going to do when you get into high school, and then what are we going to do in the two-year college and, my God, what are we going to do for the senior year of university? And I think we have to move away from that fear and move into those outer edges and expand beyond that capability that we have now.

CALCULATORS

Leitzel: Our students are calculator-dependent. There is no question about it. They bring extra batteries to their exams. They depend very heavily on that calculator. They have it with them night and day, under the pillow, whatever. They will not have this constant interaction with the microcomputer. They will come and go. That will be a disadvantage for the approach that we have been using at Ohio State. It's clear that writing programs enforces the use of variable and strengthens the knowledge of algebraic representation. But it may be that students need to have a fair idea of what a variable is before they can write short programs to assist themselves in solving problems.

Page: Can your students go from a graph to a table, from a table to a formula, from a formula to a graph?

Leitzel: Yes, yes, yes! But they often do not elect to. They often will elect to take the numerical representation or the graphical representation if they can avoid that algebraic formalism. But I think that we have asked questions that suggest that they do know that the same phenomenon, the same relationship between quantities, is being represented in each of those three.

Page: You point out that your students are very dependent on their calculators. To me that's also very dangerous; I think that it's very important to convince students that one's head-held calculator--the brain--is much more important than one's hand-held calculator. One way to convey this is the fact that you can give them problems that you can pose with a calculator that you can't solve with a calculator. In fact, in my paper, I've given a couple of those instances. What is being done to be sure that students don't become calculator junkies?

Leitzel: I think we do make an effort to convince students that their calculators would not do anything if it weren't for them--that they have power over that calculator. We do ask questions when we rewrote the placement test so that students could use calculators and it would not have a calculator bias. We wrote questions so that the solution was not numerical. We do that on our tests, too. We don't ask for the tangent of 60 degrees, or something like that. We ask for relationships within the problem. We often ask students to write in paragraph form what is going on somewhere as an intermediate step before they can do things in a more formal abstract form.

Davis: Joan, I support wholeheartedly the ideas that you have put together here. I've seen in some enlightened situations, in terms of vocational math courses, this exact style being utilized. I've seen students who were considered total failures in high school mathematics, students who start such a course with a calculator dependency, coming out confident and coming out with very good problem-solving skills. In terms of your comment, Warren, yes, they are calculator-dependent at the beginning. It's interesting to see how they become weaned from that calculator as time goes on and they gain confidence because they have a conceptual understanding now of the processes, whereas before they were pounding up against the wall where the arithmetic was getting in their way and they failed to grasp the concepts. Now the calculator assists them in grasping the concepts so that they can then go back and concentrate on the arithmetic skills, confident about the concepts that they are dealing with.

Let me add one point. I think what we're also talking about here is doing what Jim Kaput had talked about yesterday--going to a different level of looking at problems. We're developing problem-solving skills in students whereas before all that we were asking them to do was perform manipulations, and, in the long run, we're coming out with a better student from this kind of approach.

Tucker: I agree with Ron's comments. No one ever worried about engineers and their slide rules--when you have to work with numbers, you have to work with numbers.

Kaput: I'm going to follow up on that and tell Warren Page that in fact "you're an information technology junkie, because if I asked you to multiply 37 x 56, you are going to go to a certain information technology rooted in the base ten place-holder system and an algorithm for multiplying numbers; you're going to write those numbers down in a certain order and carry out that calculation; and, in fact, you are hooked on that particular information technology." It's critical to realize that information technology has been used for many years. In fact, it was an enormous advance made by human culture when the placeholder system was invented and then the algorithms that got based on that came to be known to people. We collected them and we got more capable as a result of that. Before the base ten placeholder system, very few people could multiply 37 x 56. With the use of that information technology a lot of us can do it.

Ellis: I'd like to issue a warning about the use of the word "variable." A couple of days ago I was looking at a MUMAX Manual (MUMAX is an editor used by M.I.T.) and in it the manual writer said "a variable is a place to store information." That's not the same idea that we have in mathematics; and as we go into computing, I think that's something we have to be aware of. But if you look elsewhere, you often find "a variable is a storage location."--It doesn't get any better.--Kids know about variables when they are seven or eight years old now, those who have worked with computers, and they know about variables that have other names than X. They can have names like "Bob" and "Fred" and "George." They have a richer idea of variable than used to be brought to a mathematics course, but it's a different idea and there may be some need to cope with that.

Warren: There is a statistic floating around that says that around 20 percent to 30 percent of third graders have a negative attitude towards math and science and when you get to the seventh and eight grade I think that's increased to about 80 percent. I think that the utilization of the technology that you have in your hand to provide the knowledge and content certainly is going to facilitate the learning of those students. You may be a little pessimistic in terms of the utilization of the microcomputer and the availability of it. We already have, released in the last two or three months, the wristwatch T.V. that's available at a cost, I think, of less than a thousand dollars. That's a Dick Tracy thing that I saw in the comics 25 years ago, and it's now reality.

I have 175 electronics-electrical students on my campus and every one of them has a calculator in a leather pouch and you never see them walking on campus without one.

Leitzel: Let me say a short word about computers and equity. I think that there is an issue that we have to keep tuned into even though we don't like these kinds of things. There is a possibility that the availability of computers will make a wider separation between the economically disadvantaged and the economically advantaged children as they come into our schools. It's also possible that increased use of computers will make a wider separation between the success of girls and boys and the learning of mathematics at grades six and seven. There's not a little boy in Columbus, Ohio, who isn't on a computer by the time he's eight years old. Computers are in the libraries and, if you walk through the libraries, you can see the little boys all over those computers. But that is not the case with the girls. In the seventh and eight grade classes that we worked with this year, many of the girls--and this is a suburban district--may have had computers in their homes but managed not to get very involved with them. They were "Dad's computer."

DISCRETE MATHEMATICS

Maurer: What about the students who probably are never going to learn enough to be able to do certain complicated manipulations? Is there some way to leapfrog that with the current technologies and somehow get at concepts, or is there a way of giving them other material? If we mix, let's say, some discrete math material--an easy sort, not the sort that I'm proposing for four-year college work--then at least it's new material, and they may be interested to see applications of it. Some of that can be done with or without computers and not be so difficult in terms of manipulations. That might be much more useful to them and much more fun.

Kaput: Another route to math understanding that hasn't been as widely discussed yet is a real shortcut that doesn't quite go to the same place. If you realize why you want students to know that mathematics "up there"--that is to be able to solve a certain class of problems having to do mainly with optimization and a number of other things--it is possible to do a lot of those problems using computer technology, where all the mathematics is now pushed down into the machine and out of sight.

Let me give you an example. In the calculus we often lead off the applications of derivatives with the classic canonical fence problem. Maximize the area of a pasture enclosed by the fence, and the pasture is usually along the river since a three-sided fence will make it a little more interesting, which I think is also a canonical example of a bad problem. If you had a machine available that could plot for you a graph of the appropriate area function, then you've got it. You don't have to go through the calculus to get the optimal solution.

But beyond that, if you are in fact dealing with a more realistic situation, you could use something like a spreadsheet and plug in the information. Suppose you've got a business situation, and you've got a couple of interesting or important output variables like profit and revenue, perhaps. You sort the situation out and that requires some serious thinking and some serious work. You identify several of the variables and then you begin. You put those into your spreadsheet and start putting in numbers; it's really a fancy multidimensional version of what Joan Leitzel was talking about earlier. You start putting in numbers and watching what the outputs are.

Well, number one, you've gone past single-dimensional calculus to multi-dimensional analysis, but now you've got control of a much more complex situation. You very systematically vary the input variables. Look at what their impact is on the output, and you discover that if you let the variable x run across a wide margin, it doesn't make a lot of differnce in the output, but if you change y just a little bit, and lo and behold it has a large impact on the output. You've learned something about the situation; z, on the other hand, just seems to act in a strange way. You change it by a certain amount and the output changes by a predictable amount. Well, you might say, "Gee, x is not important. Let's fix x because it doesn't change much, and to the machine you say, "Why don't you give me a plot of y and z? Give me my upward variable of profit, plotted as a function of y and z." And you look at the surface now that appears on the screen, and you say, "Ha, that's why when y changed a little bit, I had this big change in output. I can see that now on the graph because the graph has a big falloff point there."

Now, what's going on there? You are getting some very sophisticated analysis, which previously used some sophisticated mathematics; calculus of variations is not very far from what you're doing right there. Of course, you're using very primitive methods when you are using the brute force power of the computer to analyze this complex situation. So somehow you've gotten to the mathematics of calculus and variations, which very few people ever get to in their curriculum. You're accomplishing some of the same things with the computer, using the information technology and using its ability to compute to build your graph. So it may be possible to become a serious mathematician in terms of being able to solve some serious problems without having to go through all of that formal mathematics.

Renz: Jim, this is already the way people work. I mean, you're saying one might build a model like this, but this is the way business people work. They don't use calculus; they use arithmetic. And as to spreadsheets, the power of the spreadsheet for modeling various things, that's something that has captured people's minds over a rather broad spectrum of problems.

This whole problem of optimization brings me back to Steve Maurer's paper, and I would just like to read one paragraph out of it because I think this is exactly the wrong way to think of things. Steve has said: "In short, we educators are confronted with the following optimization problem. If ninety percent of the students need to know something about A, and eighty percent need to know something about B, and seventy percent need to know something about C, etc." First of all, I don't think we are confronted with an optimization problem. Second, no one has any idea what percentage of students need to know *what* about *what*.

Maurer: Let me disagree with that. We don't know what will be the most useful techniques for solving certain problems ten to fifteen years from now. We don't know what individual students are going to do with their lives, but we do know, for instance, that discrete techniques have been more useful than continuous techniques in doing economic problems of a certain sort. We do know that, say, a graph theory representation, in a graphical sense, is more helpful for getting a lot of ideas started than the graphs of functions in calculus. We can make a rough stab, and that's all we can ever do if we are deciding what are the most valuable ideas and specifics that people should know. But I think that even with that rough stab we can attempt to make a greedy algorithm solution. Given that students drop out as they go along, it's pretty clear to me that the current order of three semesters of calculus before they get to any other mathematics is not the right order.

Tucker: I think that things have to evolve down and discrete math can't come in out of nowhere; it really should have an audition in the curriculum. The four-year colleges are going to infuence the two-year colleges. At the junior level, I see a growth which shows that it's ready to move down, and I think that we should be thinking about discrete math coming in, in a natural evolutionary way.

Rodi: In a sense, it seems to me that the pure mathematics that overtook things in the fifties and the sixties had as its model this wonderful deductive learning, and you had to climb a steep path to get there. When you got there, you had this fabulous tool--calculus. With computers, spreadsheets, discrete mathematics, statistics, and exploratory data analysis, we're finding a movement back to the idea that learning originally comes inductively. Perhaps we have gotten too deductive. It now takes too long to climb the mountain, and when we get there the atmosphere is so rarefied and it's a tool that is not useful for the great mass of mountain climbers. And, besides, you sort of lose touch with the roots. As St. Thomas Aquinas said, "all knowledge ultimately comes through the senses"--it all comes through induction.

Warren: I think that Peter has put his hand on one of the frustrations that vocational/technical instructors have with mathematics departments, and that is the optimization problem is not really to be dealt with. That is, we don't know what the content is. If the vocational/technical faculty felt that way, the vacuum tube would probably still be in any curriculum in which the instructor who taught that twenty years ago was still in that institution. We derive our content from analysis and we have to determine from that what, in fact, makes up that industry or that technology so that we then can teach it. When transistors were first introduced, the instructor who was teaching vacuum tubes had to very quickly pull that in, so that whenever it showed up in circuitry it could be taught. I can't believe that the same thing cannot be done in mathematics. I'm certainly not saying that we need to turn back to consumerism or to the job-placement orientation that we've had in the past ten to fifteen years, but I do believe that the content of mathematics can be determined by doing appropriate analysis of the client, or the graduate target placement that we are dealing with in junior colleges. One of the serious problems that I have heard discussed over the past year during my sabbatical in Washington is that there are many junior colleges which have become community colleges but that the old transfer orientation still exists in the minds of that faculty and it will not change until that faculty has a complete turnover, and community college faculty are hired in place of the junior college faculty.

Evolution or Revolution

Page: A natural concommitant to the need for remedial mathematics is also the need for remedial writing and reading. What about the style of exposition? What style of exposition is needed for, say, remedial mathematics or discrete mathematics? Given the topics, it seems logical to me that there should be various levels of exposition geared to the particular needs of the students involved. I think we're losing a lot of students because they don't understand the transmission; it's not the ideas.

Renz: The textbook is a curious kind of item--like a prescription drug, it's bought by one person, but its use is dictated by another. Textbooks are written, among other things, with the idea that they will be adopted by the professors who teach the course and they are, in that sense, directed to the professors. They are also supposed to meet the educational needs of the students, and there is a double bind there.

Gordon: You mentioned that the reading and writing problems were concommitant; I don't think they are. I think they must precede the mathematics.

Rodi: One of the very tools we are talking about using is the video screen. It has probably been responsible over the last twenty years for bringing students to the point where they don't read. In a sense, we are arguing against ourselves by saying, "Let's put more of it on the screen." Yet, we all know that if you are going to be precise you have to be precise with language. For many of our students who have not taken mathematics in high school, the first time that they are required to be precise with language is in the math class.

Warren: I would say that some students who cannot understand sentences you express in your math class, Shelly, would probably immediately lose you if they tried to describe the operation of a transaxle in a new front-wheel drive automobile. But those technical faculty do have to take the time to teach the vocabulary. I'm not sure the mathematics faculty should say that they don't have time.

Part 5.

THE LEARNING OF MATHEMATICS

RESEARCH IN THE LEARNING OF MATHEMATICS: SOME GENUINELY NEW DIRECTIONS

by

James Kaput

SUMMARY

Research in how people learn and use mathematics has been changing in fundamental ways during the past decade. Increasingly it draws its organizing principles, perspectives and methodologies from the new Cognitive Science and Artificial Intelligence disciplines rather than from classical educational psychology, which traditionally has informed mathematics education research. This change, part of a larger evolution associated with the arrival of the Information Age, has led to detailed observations of college students doing increasingly realistic mathematical tasks. The resulting analyses are used to build explicit cognitive models describing the mental structures and processes involved in doing these tasks and related errors. This work provides scientific bases for the following general results.
(1) Students at all age levels have little ability to relate their school mathematics to the wider world of experience.
(2) All students bring to every learning situation primitive but remarkably robust intuitive conceptions that are the primary means by which they interpret their experience, including any school explanations or theories.
(3) Novice/Expert differences include more than quantity of information, but much more importantly, profound differences in the structure of that information and differences in the amount of informal, qualitative, imagistic, metaphoric, and heuristic knowledge that experts use (and of which they are mainly unaware).
(4) Students must construct knowledge by actively acting on and through old knowledge structures - communication metaphors about learning need to be replaced by construction metaphors.
Information technology is radically altering educational objectives as well as the means by which they may be achieved. Thus we must expand our understandings of student thinking so new curricula and pedagogy will fit the cognitive realities being uncovered. Fortunately, new software developments promise both improved means to examine the workings of students' minds as well as the means to improve them.

RESEARCH IN THE LEARNING AND USING OF MATHEMATICS: SOME GENUINELY NEW DIRECTIONS

James J. Kaput

Introduction

Those that can, do.
Those that can't, teach.
Those that can't teach, teach teachers.
Those that can't do real research, do education research.

George Bernard Shaw's words in the first two sentences, extrapolated to the second two, underline our starting point that research in mathematics education has historically had little claim on the attention, let alone respect, of the college mathematics teaching community. This paper will attempt to show that this relation between the education research and the teaching communities is ripe for real change.

The organization of this paper is as follows:

1. We preview some of the new discoveries, influences, and trends in research that will change the way mathematics education will be viewed and applied in the future.

2. We sketch some background on the general types of mathematics education research that have taken place over the years.

3. We describe actual research studies at different levels of mathematics education that have a bearing on the teaching of mathematics at the college level and that illustrate new and promising trends in research.

4. We discuss the role of mathematics education research, especially in connection with the profound impact of information technology on what and how we teach, in managing for our students' benefit the deep changes soon to affect all segments of the mathematics teaching community.

1. New Directions in Mathematics Education Research: A Preview

In the past decade a new interdisciplinary field of inquiry has developed which draws its organizing principles and tools of analysis from three areas, cognitive psychology, computer science, and artificial intelligence. This field, called Cognitive Science, is replacing traditional educational psychology as the central contemporary influence on mathematics education research. This change has been accompanied by a host of others. The most important change is in how it is done: there has been a shift away from simple "controlled variable" statistically analyzed studies toward formal and informal cognitive model building based on the close observation of individuals. The shift in methodology and aims inherent in model building reflects a corresponding shift in the conceptions of science in recent years that gives a much richer and more realistic view of the process of scientific theory development than the impoverished view of science that informed classical learning theories and educational psychology earlier in this century. The models, informal and formal (often in the form of computer simulations) play the same role played by the informal models and formalisms of the traditional "hard" sciences in guiding the search for and interpretation of data.

This development is actually part of a more profound change that has been occurring as part of a larger shift in milieu associated with the emergence of the computer and information sciences. In Newton's time there developed a mechanistic world view that accommodated a new set of entities such as force, momentum, action at a distance, and so on. That world view disregarded the distinctions, only recently sacred, between heavenly bodies and earth - both contained the same entities which obeyed the same laws. That world view went on to support the Industrial Revolution and remained relatively intact until this century. A number of causes too numerous and complex to account for here led to the gradual weakening of that overarching cultural framework. It is gradually being replaced by a new framework that is commonly associated with the phrase "The Information Age." This new framework accommodates some new entities - information, structure, and process. These are the new realities and, more important to our purposes here, they are applied across boundaries that were only recently sacred. They disregard the distinction between mind and machine. They are now assumed to be the same entities in both arenas, and the same laws obtain. In particular, what we learn about the structure and manipulation of information in machines applies to the human mind, which itself is an information processor. In a very deep sense the epistemic gestalt has been altered, as has our epistemic self portrait. From this change many others are following, including the way we approach the study of mathematical thinking and learning. The new approach is commonly referred to as the "information processing paradigm."

These historic changes, when coupled with the understandings accumulated in earlier decades, have led to insights that in turn have produced a substantial new consensus regarding what it means to learn and use mathematical and scientific concepts. They have also helped expose some of what may be missing from our current curricula. (There has also been a blurring of the distinctions between mathematics and science education research - after all, information is information.) We now sketch a few points of consensus generated by the intense research activity of the past several years.

(a) Students at all age levels, including those acknowledged as excellent students, have little ability to relate their school-based knowledge to their wider world of experience. This point is verified both by the large scale statistical information arriving from the different National Assessments of Educational Progress and by the observation of individuals who are presented with tasks calling for application of what they "know" from school mathematics and science courses to realistic situations.

(b) All students at all levels bring to <u>every</u> school learning situations primitive but exceedingly powerful and resilient conceptions that are the primary means by which they interpret the world about them, including the learning situations themselves. Such primitive conceptions (about the physical world, numbers, geometry, infinity, part-whole relationships, logic, proof, algebraic symbols, and so on) control both how students interpret the concepts and terminology they are presented in school to learn as well as how they approach non-school applications that nominally involve this school "learning." The resilience and robustness of these primitive conceptions have been one of the true surprises of recent research - they are extremely difficult, perhaps impossible, to "teach away." The relationships between these primitive conceptions and the formal concepts of mathematics and science are being mapped out, but it will be a major task to determine exactly how to alter curriculum and instruction to account for these most powerful forces in particular cases.

(c) The difference between experts and novices in any particular field includes much more than quantity of information in that field or even cognitive access to more information. Instead, the difference seems to be twofold: (i) in the organization of that information into extremely large and highly structured chunks, and (ii) in the amount of informal, qualitative, imagistic, metaphoric, and heuristic knowledge that the expert is able to bring to bear in particular situations. This latter knowledge is very weakly, if at all, represented in textbooks and formal instruction, but on the other hand appears to account for much of the performance difference between experts and novices. Indeed, detailed examination of expert performance has revealed that experts do not usually realize that they have or are using this non-textbook knowledge. If they did, perhaps it would be better represented in the curriculum.

(d) Students are not blank slates or empty vessels who passively receive knowledge; they are active constructors of knowledge, continually building new knowledge out of old knowledge (which is sometimes made up of the primitive conceptions described earlier) by acting on their experience. Moreover, the processes by which the construction of lasting ideas or reliable procedures takes place are exceedingly complex and extensive across time, and vary significantly from one subject domain to another. The failure of simple explanations or proofs, however convincing to the explainer, to build viable knowledge structures in the mind of the learner can usually be traced to an inadequate opportunity for the learner to act upon appropriately meaningful experiences ("meaningful" in terms of the student's prior conceptions.)

In later sections we shall expand on some of these generalities in more concrete terms using examples selected from current research. However, worth mentioning is one more consequence of the methodological shift toward close observations of individuals doing complex mathematical tasks. Unlike the majority of research results of classical studies in mathematics education, which were statistical statements about populations, the newer outcomes tend to be cognitive models of individuals. And these individuals are typically modelled doing tasks that are relatively close to the kinds of tasks that mathematics students normally do. For these two reasons the distance between the research and the classroom is much reduced, leading to greater and more immediate applicability of the research.

2. Status, Fatalism, and Professionalism

Let us now briefly return to the question of why mathematics education has had such low standing in the community of mathematics teachers. The answer turns out not simple and fairly far reaching in its sources. But first we can dismiss one superficial answer.

Often one hears from mathematicians the complaint about mathematics education literature that it is loaded with jargon - this from the field that brought us -- fill in your own best example here -- No, it is not jargon. Every area of disciplined inquiry requires special terminology for purposes of precision in communication and for expression of ideas, distinctions and relationships that are not readily carried by common parlance. Mathematics is surely the creator and purveyor of more jargon, especially in the form of special symbols, than any other field. And much (but surely not all) of this jargon is necessary. The <u>real</u> issue is whether the jargon is justified in any particular instance. And this question inevitably turns on whether something of significance is being expressed by that jargon. Most mathematicians believe that in the case of mathematics education, nothing of significance <u>is</u> being expressed. The roots of this belief pattern reach both the firm soil of truth as well as the soft sand of ignorance. And where there is ignorance, one can usually find unacknowledged fatalism.

A leading contemporary researcher, Robert B. Davis of the University of Illinois, in answering the question of why the study of mathematics is so old while the study of mathematical thinking is less than a century old, offers this revealing comparison of mathematics education with medicine:

> mathematics education is in this respect somewhat like medicine - centuries ago there was very little sense of control, and more acceptance of fate in both fields. If many infants died, that was what happened; only in recent years have we adopted the point of view that every single infant death must have a rational cause, that these causes must become known, and that this knowledge must be on so fundamental a level that appropriate interventions can be devised. The modern point of view has virtually eliminated polio, entirely eliminated smallpox, provided a measure of control for diabetes, and greatly reduced the impact of cardiovascular diseases. Interestingly enough, though, strong traces of ancient fatalistic attitudes still survive into late twentieth century medicine: we tend to regard the common cold as 'just one of those things,' and many diseases of old age, such as arthritis and failing vision, are commonly accepted as essentially inevitable and incurable. (Davis, 1984, p. 3)

The currently common, fatalistic view of mathematical thinking, learning and teaching very much parallels the older view of medicine. Some students can do it and others, like the infants of yore, are the victims of fate. Some mathematical ideas are hard to learn and others are easier, some mistakes are common but most are random, some teachers seem to be better than others but we must all cope as best we can, and this is merely the way it is.

And there is good reason for this fatalistic attitude. Few would argue that the insights into the workings of our students' minds provided by the bulk of research into mathematics education earlier in this century told us anything beyond what a reasonably competent teacher already knows, either explicitly or implicitly. Similarly, few curricular decisions, especially at the college level, seem to be driven by research results. Our "folk" methods have been as effective as any provided by the discipline of mathematics education. Given this state of affairs, a view of mathematics education research as a kind of ignorable "quackery" directly follows.

But why is this the case? There are two deep and intertwined root systems supporting this attitude. One has to do with matters of status and the other has to do with genuine effectiveness of the research in question. A detailed analysis of this issue will be published elsewhere.

These matters of status and effectiveness have direct bearing on the spectrum of professionalism issues raised in both the Conference Recommendations and in several papers of this volume. This author believes that, until recently, activity in the area of college level mathematics education research has had neither the status nor intellectual substance on which to build a professional community of teacher-scholars dedicated to an enterprise whose goal is the improvement of learning at the college level. Papers or conference talks on the subject of Math Lab Organization, Retention Rates and Tutoring, etc., while occasionally helpful, will not serve to crystallize the attention, focus the energy, and engender the authentic academic respect necessary to support the self-sustaining sense of professionalism now felt to be lacking among many TYC mathematics teachers. The opportunities - and challenges - now exsit for genuine scholarly activity that on one hand is squarely in the tradition of academic research in areas such as mathematics and science, and on the other, is directly applicable to the difficulties we experience daily as teachers. The following section seeks to illustrate what we mean by "research" in this sense.

3. <u>The Trend Towards Cognitive Model Building</u>

Mathematics education as a discipline in its own right began near the turn of the century, with the greatest amount of attention paid to the learning of arithmetic in young children. There were two basic reasons for this focus - one having to do with the fact that mathematics education was a spinoff of psychology and the other having to do with the relative simplicity of the subject matter, which made it easy to study. Over the years these early trends have held - most of the theory and methodology in mathematics education have been borrowed from various branches of psychology, and grade school mathematics, especially arithmetic, has been the most extensively studied subject area. One can still see divisions between research that is more oriented towards cognitive psychology, where the tasks and the theories tend to the general and the models are less detailed, and the mathematically oriented research, where the tasks are more complex closely tied to mathematical topics (Hiebert, 1984).

There has also been a broadening of research interests over the years to include much more of mathematics learning and other aspects of the educational process such as problem solving, student attitudes, social dimensions of classroom and school organization, among others.

We will now provide a quick overview of the traditional methodological approaches to mathematics education research.

(a) Survey research of relatively large populations usually tries to give a descriptive picture with respect to certain selected variables, such as student achievement with respect to a particular topic. The various National Assessments over the years are of this type - and have had considerable influence on policy over the decades. While this type of work can be the basis for considerable data analysis, its value in research has mostly been to point to areas needing further investigation.

(b) Experiment-based research is the application of the traditional methods of the laboratory sciences, whereby certain variables are carefully controlled, effects are measured, and descriptive and inferential statistics are applied to the results, often in the form of hypothesis testing. A great variety of experimental designs are, of course, possible, and these have tended to follow fashions driven in significant measure, it appears in retrospect, by the kinds of computing power available to support data analysis.

(c) Case study or clinical research involves the close study of individuals or situations without the tight control and manipulation of variables as in the classical experimental methodology. It trades control for flexibility and intensity of observation, which leads to a richness and depth of data with a corresponding loss of simplicity and, in some cases, objective generalizability of the results. It allows for data gathering in much more realistic situations than the experimental study which must usually factor out such uncontrollable variables as experimental environment (which might be a classroom with a teacher and students whose characteristics may be well beyond control), and which usually must control the type and sequencing of tasks. In fact, the teaching experiment, made popular after the publication in the United States of Krutetskii's lengthy studies in the Soviet Union based on such methodology, is a tool designed specifically to observe on an extended scale some of the hardest to control variables. There are deep divisions not only in the mathematics education community but in the social sciences more generally regarding the relative value of the experimental vs the clinical approaches, although the tide has turned more in favor of the latter in recent years with the waning of philosophical positivism and psychological behaviorism. Both sides agree, however, that clinical methods provide excellent means for generating hypotheses and questions to study, which for a discipline as young as mathematics education may be the most important activity. (The Journal for Research in Mathematics Education, the major American conduit for mathematics education research, sponsored by the National Council of Teachers of Mathematics, is barely fifteen years old as of this writing.) The question is not whether either of these approaches is better in any absolute sense, but rather, which type of methodology is appropriate to the research task at hand. Many of the larger scale research projects of recent years use combinations of methodologies.

(d) Evaluation research typically looks at the extended impact of some partially controlled variables such as learning materials or form of instruction. (A course final examination contains some of the elements of an evaluative study, although typically is much less formally constructed and analyzed.) Evaluation studies are of two types, formative or summative, according to their purpose. Formative studies are used as a means to help improve whatever is being evaluated, whereas summative studies are used when information is needed on whether a particular enterprise or product has the expected or desired impact, without regard to its being repeated or improved. As the final exam example indicates, the division between these two types is not always precise.

(e) Historical or philosophical research typically examines larger or longer term events, frequently including other forms of study as its data. For example, one might study the influence of political climate on curricular reform over an extended period, or examine the role of statistical packages on the epistemological assumptions and research methodology of mathematics education research, examine the historical development of a particular mathematical construct and whether it parallels the development of the understanding of that construct within individual students, or examine the impact of the demographic changes in postsecondary student populations occuring during the sixties and seventies on the first two years of the college curriculum.

Not occurring on most traditional lists of mathematics education research types, but playing an increasingly important role as indicated in the preview comments, is a sixth type involving explicit model building. It is in this type of research that the strict dependence on core concepts and methods of psychology is augmented by the influence of other fields such as computer science and artificial intelligence. This approach is also providing the theory building mechanisms that appear to be critical to the progress of any science and which had been lacking in much traditional mathematics education research.

(f) Cognitive model building research attempts to create representations of plausible causes of observed phenomena in much the same ways that models have been used in the natural sciences for centuries. (Note that the recognition of the key role of modeling in the development and nature of scientific knowledge is among the significant recent discoveries about the history of science mentioned earlier that is affecting the way we view mathematics and science learning itself.) The models can vary in their degree of formality, from strictly informal models described in ordinary language, perhaps supplemented by diagrams or other common forms of representation, to formal mathematical or computer based models. These models are increasingly used to describe individual mathematical performance. They are webs of explicit hypotheses, the relationships among which are subject to empirical investigation using any of the previous five types of research; although as indicated above, clinical methods tend to dominate the data collection associated with the actual model building. Formal computer modeling of this sort was pioneered by Alan Newell and Herbert Simon (1972) and is now a staple of research teams at many universities across the world where a strong analogy is drawn between information processing systems, such as computers, and the human mind. Thus computer programs are built that "solve" certain classes of mathematical problems and perhaps even "learn" to solve others from their "experience." Other researchers, such as Davis (1984), use the same fundamental information processing paradigm, which provides a rich pool of concepts and terminology with which to deal with the enormous complexity of human cognitive activity, but use it in a much less formal and more fully descriptive manner. Yet others, such as Leslie Steffe and associates at the University of Georgia (1983), build detailed models of fundamental mathematical concept development without explicit recourse to the information processing analogy.

An underlying assumption of all this model building research is that observable mathematical behavior needs to be "explained" in terms of some more fundamental layer of meaning usually related to the acts of learning, understanding, or interpreting a mathematical idea or procedure. A cognitive model REFERS to something invisible and not directly measurable in the student's head. In this way our understanding of that behavior is deepened just as our understanding of health and illness phenomena are deepended by the layers of meanings provided by the science of medicine and associated fields such as the biology of organisms - although in the latter case there is much more measurability and observability. Mathematics education research has much more in common with, say, nuclear physics, where model phenomena that ARE observable and measurable. For example, to discuss nuclear "particles" is already to utilize a model. And, of course, the constituents and relationships among such "particles" are only discussable through the use of some very fancy mathematical models. For a fuller discussion see (Kaput, 1984).

We shall see in the next section how this modeling approach is being applied to yield new insights into our students' mathematical behavior. But before finishing our characterization of mathematics education research methodologies, we should point out one other general strategy that has become quite prevalent in recent years, the close examination of errors as a means of gaining insight into otherwise very difficult to identify processes and structures. We shall see several examples of this type of analysis in the next section, but it may be worth pointing out that error based studies in fact lie in a long and distinguished lineage of scientific strategies. But the process of observing students making errors while working at the boundaries of their competence shares a motif with, for example, nuclear physics - one gains understanding of a system by perturbing it and then observing the results, especially when the perturbation excites the system out of its standard orbit. Another analogy can be drawn with brain research which examines neurologically damaged brains for clues regarding the organization of brain function.

4. Examples of Mathematics Education Research

Since research topics include virtually any aspect of the teaching and learning of mathematics at any level, we must necessarily be selective, ignoring such active areas as problem solving (especially the critical matter of transfer of problem solving skill), individual differences (especially spatial and sex-related ones), brain organization and neurological issues, the functional role of mental imagery, among many other areas. Kilpatrick and Wagner (1983) provide an excellent entry point to the literature. We chose three areas that either deal with mathematics of interest to a post-secondary audience or illustrate well the direction and style of future research, especially as informed by cognitive science. Our first example illustrates research that was intended to illuminate matters of importance to one age level but sheds significant light on all other areas.

(a) Early number-concept development

Before there can be arithmetic, there needs to be number. And before there can be number, there needs to be a world of objects and substances that can be counted and measured. The major twentieth century force in discovering the logical and psychological precursers to the ideas underlying those of elementary school mathematics was Jean Piaget (Gruber & Voneche, 1977). While not directly concerned with school mathematics and teaching, his influence on mathematics education research has been considerable in the last twenty five years. (He regarded himself as a "genetic epistemologist," one who attempts to find the sources and regularities of human knowing within the biologically driven unfolding of human thought - from earliest infancy onward.) His close clinical observation of young children fifty years ago led to the profound discoveries that not only is the idea of number actively and gradually built up in the mind out of experience during the first four or so years of life, but even the objects that might later be counted or classified need to be CONSTRUCTED from one's experience. Whereas mathematicians (and just about every one else for that matter) took the world of permanent objects and even the counting numbers as givens (Kronecker's statement about the god-given nature of the positive integers comes to mind), Piaget taught us that the basic ideas of number are founded not only on one to one correspondences between sets, but have a much more complex base reaching into the construction of the elements of the sets themselves.

Thus we now realize the deeply constructive assumption that must be made about the learning process - the communication metaphor that takes knowledge as fixed and somehow platonically "out there," to be transmitted from one who possesses it to one who doesn't, turns out to be wrong. Instead,

Piaget's early investigations, followed by decades of detailed study of the later cognitive development of children up to adulthood on many continents, have taught us that ideas must be constructed out of action on appropriate experience, physical and visible in the early years and more abstract in later years. Another highly relevant lesson from the work of Piaget and his followers is that student versions of ideas, especially in the absence of that appropriate experience, can be profoundly different from adult versions of ideas bearing the same names and even the same verbal descriptions. The implications for education are far reaching, for they tell us that explanations alone do not cause the necessary mental constructions to take place - unless the "explainee" happens to be a trained mathematician for whom such explanations _do_ provide the appropriate experience on which (s)he can act to construct the knowledge being explained. They also help to undermine the clean distinction between knowledge production (pure research) and knowledge communication (teaching) because _everyone's_ real knowledge must be constructed.

Extended clinical work by Steffe and associates (1983) as well as others has revealed the extraordinary richness of the genesis of the number concept via extremely close observation of students over a period of years, revealing previously unnoticed subtle regularities in children's counting behavior. This and related work has shown that just as permanent objects need to be constructed by individuals over a period of time, numbers, as objects or units, need to be constructed as well before arithmetic can be reliably learned. After all, one way to describe arithmetic is as the manipulations of such objects. Further work has recently shown that the notions of multiplication and the place holder system likewise appear to be based on the ability to unify larger collections of numbers into units. In each case we see the process by which working with collections in certain ways eventually transforms them into entities, which can then participate as objects in larger constructions - in much the same way as algebraic expressions are at first compact descriptions of numerical relationships but must then become objects to be manipulated in their own right. Similarly, functions, which begin as transformations, in turn become objects to be added, multiplied or even operated upon by derivatives - which themselves later are treated as objects (operators). The point of this discussion is to highlight the continuities being found in the ways cognitive processes are involved in the mathematics that we teach at all levels and to illustrate the detailed nature of many of the more recent investigations.

(b) Algebra manipulation skills (including equation solving):

Much of what has been done in this area seems to fall in the category of research that ratifies what most teachers would expect. As mentioned earlier, such research has yet to develop a strong and comprehensive enough theory to tell us much we do not already know. Two promising avenues of inquiry, each tied closely to the information processing point of view, include first expert/novice studies as exemplified in Lewis (1981) based on a combination of the pioneering Artificial Intelligence work by Bundy (1975) and a general approach to skill development outlined by Anderson (1981). The second is the deep/surface structure analysis done by Matz (1980) designed to explain both the errors that are commonly made as well as the extrapolation techniques by which algebraic knowledge is used to solve problems that have never been previously encountered.

Lewis uses detailed comparison of algebra manipulation strategies and parsing activities done by novices and experts on the same problems to uncover precisely what the experts know that the novices don't, especially in the case that the novices are already able to do the elementary operations reliably. The key word in the previous sentence is "precisely," and this is where the AI influence is felt. The difference should be specifiable enough to be implemented in an explicit program that, in its novice version "solves" algebra problems using the same steps that the novices do, and in the expert version mimics what the experts do. The findings of this research (still in its early stages) suggest that experts have organized some of their procedural knowledge into larger chunks that they come to treat or apply as elementary operations, producing efficiencies not available to novices. Additionally, they have means of managing procedures that do not appear to be algebra-specific. Similar AI based work has been done by Neves (1978).

The work exemplified by Matz examines student error behavior in detail with a view to building a generative computational theory that accounts in a systematic way for as many as possible of the common errors that students make as part of the process of extending or applying their knowledge base to solve new problems. (See Table 1 for a list of the types of common errors that this theory can account for.)

TABLE 1

Algebra Errors Covered by the Proposed Theory

1. Evaluating 4X when X=6 as 46 or as 46X.
2. Evaluating XY when X=-3 and Y=-5 as -8.
3. a) Evaluating 2(-3) as -1.
   b) Evaluating $(-1)^3$ as -3.
4. Parsing $3R^2$ as $3 + R^2$ or $(3R)^2$.
5. Simplifying 3 + 23(S-4) to 26(S-4).
6. Simplifying 3XY + 4XZ to 7XYZ.
7. Claiming one can't multiply by X because "you don't know what X is."
8. Computing $\frac{2X}{2X}$ to be 0.
9. Computing $A\frac{1}{A}$ to be 0.
10. Computing 0 * A to be A. (By the asterisk we mean some notation intended to indicate multiplication.)
11. Computing $\sqrt{A+B}$ to be $\sqrt{A} + \sqrt{B}$.
12. Computing $(A + B)^2$ to be $A^2 + B^2$.
13. Computing A(BC) to be AB * AC.
14. Computing $\frac{A}{B + C}$ to be $\frac{A}{B} + \frac{A}{C}$.
15. Computing $\frac{A + B}{C + D}$ to be $\frac{A}{C} + \frac{B}{D}$.
16. Computing $2^a + b$ to be $2^a + 2^b$.
17. Computing 2^{ab} to be $2^a 2^b$.
18. Simplifying $\frac{AX + BY}{X + Y}$ to be A + B.
19. Simplifying $\frac{X}{2X + Y}$ to $\frac{1}{2 + Y}$.
20. Simplifying $\frac{X + 3Z}{2X + Y}$ to $\frac{3Z}{2 + Y}$
21. Simplifying $\frac{X - 3}{2X}$ to $\frac{-3}{2}$.
22. Simplifying $\frac{X^2 + 2XY + Y^2}{x^2 + y^2}$ to 2XY
23. a) Computing 2(X + 3) to be 2X + 3.
    b) Computing -(3X - W) to be -3X - W.
24. Computing (AX + B)(CX + D) to be $ACX^2 + BD$.
25. Solving for X in terms of itself: e.g., getting $X = \frac{(4Z + X^2)}{7}$.
26. Factoring $X^2 + \frac{5}{6}X + \frac{1}{6}$ as $X(X + \frac{5}{6}) + \frac{1}{6}$.

Task	Typical Wrong Responses
27. Solve for X: $\frac{X+1}{X+4} = \frac{5}{6}$	X=4,2

TABLE 1 CONT'D.

Task	Typical Wrong Responses
28. Solve for X: $2X+5=11$	$X+5=\frac{11}{2}$
29. Solve for X: $3X+5=Y+3$	$X+5=Y$
30. Solve for R: $\frac{1}{R}=\frac{1}{R_1}+\frac{1}{R_2}+\frac{1}{R_3}$	$R=R_1+R_2+R_3$
31. Solve for X: $\frac{1}{X}+\frac{1}{X^2}=\frac{3}{X^2}+6X^2$	$X+1=3+6X^2$
32. Solve for X: $(X-5)(X-7)=3$	$(X-5)=3$ or $(X-7)=3$ $X=8$ or $X=10$
33. Solve for X: $\frac{5}{2-X}+\frac{5}{2-X}=4$	$5(2-X)+5(2-X)=4$

The theory begins with the following two simply stated extrapolation mechanisms for generating algebra errors which are then successively refined and recombined in the style of an AI formulation:

(i) Use of a known rule as is in a new situation where it is inappropriate.

(ii) Incorrectly adapting a known rule so that it can be used to solve a new problem.

These mechanisms are operationalized as errors in using very general cognitive processes which include the following: several forms of generalization, including generalization based on prototype problems and generalization via repetition of known rules, pattern matching (including problem preprocessing to fit known rules) and linearity, including generalized distribution rules. To be concrete let us look at the correct and incorrect applications of linearity listed in Table 2.

TABLE 2

Correct and Incorrect Examples of "Linearity" Applied to Various Rule Patterns

Correct

$A(B+C)=AB+AC$

$A(B-C)=AB-AC$

$\frac{1}{A}(B+C)=\frac{1}{A}(B)+\frac{1}{A}(C)$ equivalently, $\frac{B+C}{A}=\frac{B}{A}+\frac{C}{A}$

$(AB)^2=A^2B^2$ more generally, $(AB)^n=A^nB^n$

$\sqrt{AB}=\sqrt{A}*\sqrt{B}$ more generally, $(AB)^n=(A)^n(B)^n$

TABLE 2 CONT'D

Incorrect

$\sqrt{A+B} \Rightarrow \sqrt{A} + \sqrt{B}$

$(A+B)^2 \Rightarrow A^2+B^2$

$A(BC) \Rightarrow AB * AC$

$\frac{A}{B+C} \Rightarrow \frac{A}{B} + \frac{A}{C}$

$2^{a+b} \Rightarrow 2^a + 2^b$

$2^{ab} \Rightarrow 2^a 2^b$

Each of these is a result of applying one of the following three schemes, where $\square$ and $\triangle$ denote unspecified operators, either binary or unary. For one binary operator ($\triangle$) and one unary operator ($\square$) we have

SCHEME 1: $\square(X \triangle Y) \Rightarrow \square X \triangle \square Y$

For two binary operators, we have

SCHEME 2: $(X \square Y) \triangle Z \Rightarrow (X \square Z) \triangle (Y \square Z)$

SCHEME 3: $(X \square (Y \triangle Z) \Rightarrow (X \square Y) \triangle (X \square Z)$

In Table 3 are specific computational instantiations of these schemes, some correct and others incorrect. ("Times," of course, is a convenient way of stating "multiplication" in a binary operational form which does not priviledge either factor.) The reader is invited to derive some of the common fraction manipulation errors from the above schemes using division as one of the operators. There is not the space to specify the details of the theory, which is itself still growing, although it can be said that it does account for a surprising amount of mathematical behavior with a fairly economical set of generating principles - and it does so in explicit ways which, using AI methods, can formally derive such behavior in a computational environment such as LISP.

(c) The applications of Algebra (including word problems):
This second area of algebra research relates to student ability to connect the manipulative algebra knowledge to the wider world of experience, especially in the ability to solve traditional word problems. Again, although the lack of a comprehensive theoretical framework for this inquiry over the decades has limited the insights available from the bulk of this work, recent developments indicate a change may be in the offing.

TABLE 3

Obtaining Specific Examples from Schemas

	□ OPR	△ OPR	RESULT
SCH3:	times[†]	Plus	$A(B+C)=AB+AC$
SCH1:	Square root	times	$\sqrt{AB} = \sqrt{A} * \sqrt{B}$
SCH1:	Square root	plus	$\sqrt{A+B} \nRightarrow \sqrt{A} + \sqrt{B}$

X□Y is the base; Z is the exponent

SCH3:	times	exponentiate	$(AB)^n \Rightarrow A^n B^n$

X is the base, Y△Z is the exponent

SCH3:	exponentiate	times	$2^{ab} \Rightarrow 2^a 2^b$

[†]Since the multiplication in the distributive law is written implicitly using concatenation rather than explicitly with the binary operator "*", the distributive law may look more like an instance of SCH1 than SCH3. Using SCH1 would require the □ operator to be "*A", as if SCH3 had been curried once to yield SCH1.

One line of analysis that we shall pursue here has dealt with the deeper student understanding and misunderstanding of some of the basic ideas of algebra that are used to represent quantitative relationships mathematically, especially the critical idea of variable (Clement, et al, 1979; Clement, 1982; Lochhead, 1983; Kaput & Sims-Knight, 1983.) The studies cited begin with the finding that a majority of college freshman calculus students cannot correctly solve the following problem:

"Use the letters S and P to write an algebraic equation that represents the following statement. At a certain university for every six students there is one professor." (Earlier versions had also used the phrase "six times as many students as professors" with no statistically significant difference in performance.)

Approximately two thirds of the errors have the form of the reversed equation "6S = P". Early followup studies showed that for college age students the choice of literals had little impact on performance, nor did hints designed to reduce "careless" errors. (Performance on such tasks drops off considerably when non-integral ratios are used, e.g., "For every two hotdogs sold there are five beers sold...") Extensive clinical interviews and survey testing indicate that there are two fundamental cognitive processes leading to the same reversal, one involving an elementary word order match strategy, and the other an imagistic strategy, whereby the students create a (relatively abstract) visual image of two sets of people, one with six elements and the other with one, and then "read off" the equation as a description of the image. In each strategy, however, the letters "S" and "P" are used as natural language nouns rather than as mathematical variables, the number "6" is used as an adjective rather than as a coefficient-multiplier, and the adjacency is the usual modifier-adjacency expected between adjective and noun rather than adjacency indicating multiplication. In effect, the syntax and rules of reference of algebra are overridden by those of natural language.

None of this would be especially shocking or important if such errors were easy to correct or if the further probing did not reveal that the majority of students have grossly inadequate and unstable understanding of variables in algebra - a major impediment to learning the ideas of Calculus later. Moreover, further investigation showed similar deficiencies in performance on tasks involving translations between a variety of different representational systems (e.g., pictures and equations, equations to natural language, graphs and natural language, etc.) not easily attributable to surface, verbal misunderstandings. Lastly, these error patterns were shown to coexist within individuals, often exposed by contradictory outcomes in the same interview session, with successful performance on many school tasks usually associated with correct understanding of the appropriate mathematical ideas. In short, as with certain simultaneous investigations in science education (McCloskey, et al, 1980; diSessa, 1982), these studies uncovered a genuine mess lurking just below the surface of the everyday classroom phenomena. Exploring the roots of these errors (Kaput & Sims-Knight, 1983) takes us back to such fundamental issues as how multiplication is taught in grade school (almost exclusively as repeated addition) and how algebra is introduced (as generalized arithmetic), where each is seen as providing an inadequate foundation for the concepts that must be built on them. This investigation has also opened a theoretical window into such hard to probe questions as the role of imagery in student thinking about quantitative relationships (Sims-Knight & Kaput, 1983).

The objectives of the above types of error-based research are twofold - one to determine sources of errors so that these sources can be eliminated by appropriate changes in instruction rather than the errors themselves (for the same reasons one weeds a garden by pulling up the weeds by their roots rather than by cutting off only their visible parts.) The other objective, in common with most error-based research, is to gain insights into the organization of the knowledge in the student's mind in a way analogous to how "Freudian slips" allegedly reveal what the person was really thinking.

5. New Directions for Research in Mathematics Education

(a) A Statement of the Problem

The several progressive trends outlined in the preview section of this paper are likely to accelerate. At this point, however, one of the critical difficulties in mathematics education research is how to understand the enormously complex data that the clinical approaches of recent years have uncovered, especially with regard to the data associated with the more complex mathematical tasks of interest at the postsecondary level. The advice of Karmilloff-Smith and Inhelder (1975) seems entirely appropriate: "If you want to get ahead, get a theory." But that's the hard part. It is also the place where newer developments in information science, cognitive science, artificial intelligence, all in combination with the relatively atheoretical insights provided by the past seventy five or so years of traditional mathematics education research, can help. As perhaps exemplified best by the work of Davis and associates (Davis,

1984), these new fields are providing the basic vocabulary and organizing concepts out of which a theory may be fashioned - whether or not formal information processing models are used to instantiate that theory. However, there is far from uniform acceptance of this idea (Kaput, in press) and there are many shades of acceptance of the information processing view of the mind in general or mathematical thinking in particular, ranging from literal acceptance of the mind as a computing device (albeit not necessarily a serial, von Neumann machine) to downright rejection of the usefulness of computational metaphors by some traditionalists.

So the major developments in mathematics education are yet to come, and it seems increasingly likely that they will be spawned by needs developing out of the information technology revolution now underway. The implications of this revolution are only slowly unfolding, but with each new technological development the realization intensifies that, in Simon's words, this is an innovation the magnitude of which is to be measured on a scale several centuries in duration. As with any innovation, the first steps are to do the old things in new and perhaps better ways. However, this particular innovation is simultaneously undermining the validity of many of the old objectives of mathematics education, those having to do with computational skill at all levels, while enhancing the importance of other of the traditional objectives - those difficult to achieve objectives having to do with higher level understanding of concepts and procedures. It is also creating new objectives relating to the management of complexity, choosing appropriate representations of information, understanding graphical representations, as well as a host of more mathematically specific objectives connected with computer science, programming, and associated discrete mathematics. And if this were not enough, it is changing appropriate pedagogy with respect to almost all objectives, old and new.

The changes in mathematics topics issues are well discussed in The Future of College Mathematics edited by Ralston and Young (1983) and will not be recited here. Two other publications now widely available, Computers in Mathematics Education (Hanson, 1984) and especially the short paperback edited by Fey, Computing and Mathematics (1984) deal with some of the other issues mentioned. We will concentrate here on an examination of research results and agendas intended to lead to informed judgements regarding how to manage this enormous change for the benefit of our students.

The critical question, asked by Henry Pollak of the Conference Board of the Mathematical Sciences, is:

What is still fundamental and what is not?

The standard approach to this question, especially by mathematicians, is to examine applications of the mathematical sciences - what kinds of applications of mathematical knowledge will our students require in the future and what kinds of thinking skills will these require? The next question usually is How can we redesign the curriculum to deliver that knowledge and those skills? And finally, How can we train teachers to teach that stuff? These are all necessary questions. But there is one more that has been absent from the universe of discourse, perhaps because we have not been ready to ask it. (Recalling our early medicine metaphor, we may be like the doctors before Pasteur whose understandings and attitudes about disease did not allow a whole spectrum of questions whose answers we now take for granted.)

WHAT COGNITIVE STRUCTURES AND PROCESSES UNDERLIE UNDERSTANDING AND APPLYING THE MATHEMATICS WE DO WANT STUDENTS TO LEARN?

We intend to apply radical surgery and treatment to the curriculum - and through the curriculum to the minds of our students - without systematic understanding of how the knowledge and skills affected are organized in their minds!

Unless and until we get down to a more systematic look at the patient, we are doomed to a continual, predictable cycle of disasterous mathematics instruction, spates of national handwringing, followed by curricular revision that is well meaning, but uninformed by scientific understanding at the critical point - the mind of the student. The next subsections get down to cases.

(b) Algebra, Precalculus and Calculus

Our earlier review of research on algebra learning, both manipulative and applied, now may on the surface appear to be irrelevant, especially that part relating to algebra manipulation skills. After all, with symbolic manipulation available on an inexpensive portable personal computer, who needs to have much symbolic manipulation skill? And surely, if we don't need to teach it, we don't need research on how to teach it! But this turns out to be as wrong as wrong can be. The symbolic manipulator can solve the equation, differentiate or antidifferentiate the function, simplify the expression, and so on, with greater facility than a professional mathematician. But it cannot decide when to do it. And once the answer is in hand, it cannot decide what to do with that either! These are inevitable user responsibilities. And just as using a table of integrals required some knowledge of antiderivatives, such as elementary substitution, the user of the symbolic manipulator will require some knowledge of algebra. But which algebra and how much?

Here is where the surgery is dangerous, because we now need to know how the knowledge and skills are interrelated and develop IN THE MIND OF THE LEARNER. Matz's work has begun to show us how certain root ideas such as linearity slice across many different manipulations (although it is important to note that she is dealing with a cognitive notion and not a narrowly mathematical notion of linearity.) Using the understanding developed by this type of study, and many others of course, we can begin to make informed decisions about what to leave in and what to leave out. Similar considerations apply to the grafting onto the curriculum of new topics designed to support new applications and new thinking skills. It is imperative to know how they relate to one another <u>in the mind of the student</u>. Nor is it enough to know how they are logically related as formalisms in a formal system or how they might be related in the minds of professional mathematicians. Research cited earlier makes the point quite unequivocally that expert knowledge is very different from novice knowledge - in ways that the experts are not themselves aware.

The longer term curricular ramifications require understanding of such matters. For example, few would doubt that students' major algebra practice occurs while learning the techniques of calculus. But if these are curtailed and the time for learning preliminary algebra skills is likewise cut back, what will be the consequences on the kinds of <u>conceptual</u> understandings of algebra that support intelligent use of the technology? These questions lead directly to a whole other class of issues needing investigation.

(c) <u>Representations and Symbol Systems</u>.

We have as yet no firm grip on the relationships between "opaque" symbol manipulation skill (where the symbols are manipulated using their syntax without regard to what they stand for) and the conceptual understanding of the meanings of those symbols - how does one affect the other? And what of "transparent" symbol manipulation, where the use is primarily guided by the referents of the symbols? These questions apply across all age levels, from arithmetic to advanced calculus and beyond. In fact to date there has been little sustained research on the nature of symbol systems, or representation systems more generally. What makes a representation system carry one type of information well and not another? What does it mean to say a given symbol system "encodes" certain information? Which representations fit together and which fail to fit? And why?

To concretize the last questions, recall the infiltration of natural language into algebra, where the "Students-Professors Problem" exposed a fundamental misunderstanding of variables carried by many students. Here the given quantitative relationship was expressed in natural language as "for every six students there is one professor," and the correct algebraic equation is "6P = S". But if one uses the "S" and "P" as units, as when we write "6g" for "six grams," then one produces the common error, the reversed equation "6S = P". Clearly, we have two symbol systems interfering and competing with one another. In this case, given the weak and unstable student concept of variable, the natural language-label system usually wins (and the student loses.)

How is the representation system conflict that was described in the previous paragraph different from the <u>information</u> conflict when students attempt to understand the relation between a function and its rate of change (or slope) function? In the latter case the representation systems are the same (the algebraic and/or the graphical systems) whereas the information carried is different. Our ignorance on these matters is immense.

We shall end this paper on a more optimistic note based on the new approaches to long standing curricular problems made possible by the powerful computers now becoming available.

6. New Opportunities Offered by Information Technology

One perennial problem concerns student inability to connect mathematical ideas across representational modes, especially between algebraic and graphical modes. The fact that the parabolic graph of a certain quadratic function crosses the horizontal axis and the fact that a certain quadratic equation has solutions seem too often to reside in different sections of the student's brain. Simple and highly effective software exists (Dugdale, 1982) to connect these different expressions of the same idea in the form of an enjoyable game that basically asks students to type in simple equations whose graphs are to hit fixed, randomly located disks on the (coordinatized) screen. On one hand the constraints and scoring system provide basic motivation while on the other the computer plots points quickly and painlessly. Together, this simple combination compresses an enormous amount of intense experience relating graphs and algebra into a relatively short amount of time with the result that students quickly become astonishingly adept at predicting which equations have which graphs, which points they will pass through, how to raise or lower a graph, shift it left or right, amplify its "wiggles," and so on. Other software exists to move the other way, from graphs to equations, while other commercially available software exists to make the same kinds of translations in trigonometry. This type of quite modest, yet extremely effective software, is only the beginning of what is possible. It may also allow for the movement of some subject matter, such as trigonometry, downward in grade levels by eliminating some of the more difficult and tedious calculations. It will certainly make room for a more conceptual approach to such subjects as calculus and statistics (Fey, 1984).

Another major and perhaps twofold change is the role of geometry in the core curriculum. Despite - or perhaps because of - its longevity, Euclidean geometry has been a curricular disaster area. It has been well established that, as typically taught, not only does it fail to teach anything significant or transferable about reasoning or mathematical thinking, it also fails to add much to student understanding of space. New and emerging software makes possible an approach to standard geometric ideas which is much more active and experiential than the carefully controlled museum trip that is the form of most geometry courses. For example, the "Geometric Supposer" system being developed at the new Educational Technology Center at the Harvard University School of Education amounts to an electronic straightedge and compass which allows the student to do virtually any of the standard geometric constructions easily and accurately. The system then stores this construction as a procedure which can then be repeated on other geometric objects as well as concatenated with others to form ever fancier constructions. For example, suppose a student constructed the three medians of a given triangle (either of the student's choice or generated randomly.) The immediate question arises - was the fact that they all intersected in a single point an accident of the given triangle or is it something true in more generality, and if so, how general might be the result? Thus the student runs some experiments involving that construction on a (presumably carefully considered) collection of triangles and, if the result seems to hold up, a proof is in order. In this environment students are quite literally turned into active and mathematical geometers, often going well beyond the bounds of the standard text because (i) they are not constrained by somebody else's list of indubitable products, and (ii) they have a technological aid that supports much more sophisticated geometric constructions than they would normally be expected to do. One hopes for such mathematically rich software in other domains.

A second geometry related change of potentially larger implications involves a substantial shift in the kind of geometry done at both the secondary and college levels. Patrick Thompson at San Diego State University and others (Shigalis, 1982) have been developing microcomputer software that engages the student in active geometric tranformations making it quite clear that microcomputers are the natural medium in which to learn transformation geometry. A larger step away from existing curricula is provided by the higher dimensional geometry experiences made possible, especially on 32-bit architecture, that allow students to explore three dimensional objects and embedded surfaces by actively manipulating and transforming them. Another attractive alternative is made possible by the differential approach to geometry inherent in the Logo environment (Papert, 1980; Abelson & diSessa, 1981; Francis, 1983) which provides a stimulating, student oriented, and more direct route to fundamental ideas of higher mathematics.

All these examples indicate a "geometrization" of the curriculum made possible by information technology graphics capabilities, including parts not now treated as especially geometrical, such as calculus and differential equations. In these latter areas, most mathematicians understand their geometric side and have a few prototype examples laboriously or inaccurately produced in classroom situations which are then perhaps mimicked by students for homework. The technology encourages full exploration and exploitation of this geometric side in ways that capitalize on and then greatly expand the student's natural spatial abilities, developed over a lifetime of moving about in space. It may therefore radically alter all that has been suggested about the role of spatial ability in doing mathematics by removing what may have been a fairly artificial barrier to entry into vast areas of mathematics, a barrier supported only by the historically limited media we previously had available in which to represent that mathematics.

The role of information technology in creating application oriented environment is perhaps well enough known so that we need not dwell on it here except to note that besides the kinds of "toy" applications that are now common (including some that use real data) in rendering the mathematics of our current courses more interesting and useful, newer pedagogical uses of applications, especially simulations, will provide even more extended and fruitful learning environments. The newer environments will not only provide mathematically and data rich situations, but they will also provide for more flexible and fruitful interactions between the student and the system so that exploration and discovery can proceed more efficiently. One aid in this direction will be the incorporation of increasingly sophisticated coaches or tutors built into the system. However, the creation of such systems require both a good theory of the subject matter as well as a good theory of the student and the learning process. Again, such will only be available through research on the cognitive processes involved in learning and using the subject matter. However, the consequences of large scale applications-oriented environments have the potential for changing the pedagogical relationship between mathematics and its applications. We will be able to arrange the environments so that (i) the mathematics is needed to make sense of the phenomena and (ii) the system helps the student to generate the necessary mathematics. This turns the tables right side up finally, so that the mathematics is generated as it originally was - in response to an experienced need, rather than having the application follow as a prepackaged "word problem application" afterthought, in the feeble hope that it might work retroactively to make the prepackaged mathematics seem meaningful and worth the pain of memorizing it. Furthermore, the <u>kinds</u> of systematic rational endeavor - which we may or may not choose to call "mathematics" - will likely be expanded well beyond what now is common, to encompass ever wider domains of human experience than the rather scanty sliver now easily amenable to quantitative methods. Perhaps our students will get a real taste of that rich experience that we associate with the doing or the using of mathematics.

One last arena that we shall mention that seems ripe for significant change is problem solving and thinking in general. On one hand, there are now being developed at such places as Xerox PARC in Palo Alto the means by which students can monitor and thereby improve their problem solving strategies. The software tracks the student's progress on a mathematical problem and then processes the student interaction to produce a decision tree revealing in an explicit, reified way, the student's problem solving process which can then be examined and discussed in detail, complete with explorations of paths not previously taken. Anyone who teaches problem solving knows that among the hardest things to get students to realize is that there <u>is</u> a process, there <u>are</u> strategies, that some are better than others, and so on. This difficulty is rooted in the fundamental fact that the process is transitory, leaves no systematic record, and is thus dominated by the product, which was the presumed goal of the exercise in the first place. The technology provides means by which this difficulty can be overcome. It is but a preliminary example of the way the technology can mirror externally the processes of the mind so that they can then be examined and improved - truly a remarkable step in coming to "Know thyself!" Here we see an instance of how the technology not only promises quantum leaps forward in our ability to enable students to learn, but it also provides means by which we - and they - may study that learning to improve it even further.

So there IS reason for optimism.

References

Abelson, H. & diSessa, A., Turtle geometry. Cambridge, Mass.: MIT Press, 1981.

Anderson, J., (Ed.), Cognitive skills and their acquisition. Hillsdale, New Jersey: Lawrence Erlbaum Associates, 1981.

Bundy, A., Analysing mathematical proofs. DAI Research Report No. 2, Department of Artificial Intelligence, University of Edinburgh, 1975.

Clement, J., Algebra word problem solutions: Thought processes underlying a common misconception. Journal for Research in Mathematics Education. 1982, 13, 16-30.

Clement, J., Lochhead, J. & Monk, G. M., Translation difficulties in learning mathematics. American Mathematical Monthly. 1981, 88, 286-290.

Davis, R. B., Learning mathematics: The cognitive science approach to mathematics education. London: Croom Helm, 1984.

diSessa, A., Unlearning Aristotelian physics: A study of knowledge-based learning. Cognitive Science, 1982, 6, (1), 37-75.

Dugdale, S., Green globs: A microcomputer application for graphing equations. The Mathematics Teacher, 1982, 75, 208-214.

Fey, J., (Ed.), Computing & Mathematics. Reston, Virginia: National Council of Teachers of Mathematics, 1984.

Francis, G., Review of Turtle Geometry, by Abelson & diSessa. American Mathematical Monthly, 1983, 90, 412-415.

Gruber, H. E. & Voneche, J. J., The essential Piaget. New York: Basic Books, 1977.

Hanson, V., (Ed.), Computers in mathematics education. (National Council of Teachers of Mathematics, 1984.

Hiebert, J., Complementary perspectives. Journal for Research in Mathematics Education, 1984, 15, (3), 229-234.

Kaput, J. & Sims-Knight, J., Errors in translations to algebraic equations: Roots and implications. In M. Behr & G. Bright, (Eds.), Mathematics learning problems of the postsecondary student. Focus on leanring problems in mathematics, 1983.

Kaput, J., Representation and problem solving: Some methodological issues. In E. Silver, (Ed.), New Directions in Problem Solving Research. Philadelphia: Franklin Institute Press, 1984.

Karmiloff-Smith, A. & Inhelder, B., "If you want to get ahead, get a theory." Cognition, 1975, 3, 195-212.

Lewis, C., Skill in algebra. In J. Anderson, (Ed.), Cognitive skills and their acquisition. Hillsdale, New Jersey: Lawrence Erlbaum Associates, 1981.

Lochhead, J., The mathematical needs of students in the physical sciences. In A. Ralston & G. Young, (Eds.), The future of college mathematics. New York: Springer-Verlag, 1983.

References

Matz, M., Towards a computational theory of algebraic competence. Journal of Mathematical Behavior, 1980, 3, (1), 93-166.

McCloskey, M., Caramazza, A., & Green, B., Curvilinear motion in the absence of external forces: Naive beliefs about the motion of objects. Science, 1980, 210, 1139-41.

Neves, D. M., A computer program that learns algebraic procedures by examining examples and working problems in a textbook. Proceedings of the Second National Conference of the Canadian Society for Computational Studies of Intelligence, 1978, 191-195.

Newell, A. & Simon, H. A., Human problem solving. Englewood Cliffs, N.J.: Prentice-Hall, 1972.

Papert, S., Mindstorms: Children computers and powerful ideas. New York: Basic Books, 1980.

Ralston, A. & Young, G., (Eds.), The future of college mathematics. New York: Springer-verlag, 1983.

Schigalis, T., Geometric transformations on a microcomputer. The Mathematics Teacher, 1982, 75, 16-19.

Sims-Knight, J. & Kaput, J., Exploring difficulties in transforming between natural language and image based representations and abstract symbol systems of mathematics. In D. Rogers & J. Sloboda, (Eds.), The acquisition of symbolic skills, New York: Plenum Press, 1983.

Steffe, L. P., von Glasersfeld, E., Richards, J., & Cobb, P., Children's counting types - Philosophy, theory and application. New York: Praeger, 1983.

(DISCUSSION BEGINS ON P. 367.)

KNOWLEDGE TRANSMISSION AND ACQUISITION: COGNITIVE AND AFFECTIVE CONSIDERATIONS

by

Warren Page

Knowledge Transmission and Acquisition: Cognitive and Affective Considerations

Warren Page
New York City Technical College
Brooklyn, New York 11201

Introduction

College mathematics education will be further challenged in the 1980's by demographic changes, the enrollment of nontraditional students (older people, for example), and by society's inevitable demands for increased mathematical knowledge and competence. Thus, new initiatives for teaching widely differing student populations must be found and explored. The mathematical knowledge bases required to meet the scientific, technical, vocational, cultural, and functional needs of such varied student populations must also be closely examined. But without proper focus NOW, our approaches to these issues will be inadequate and the benefits will be ephemeral.

What is needed now is a revolution in intellectual, philosophical, and social perspectives - perspectives which reflect the very dramatic changing nature of the mathematical enterprise. Indeed, it is my belief that:

(I) College mathematics education, and in particular the mathematics training of TEMP* - career students, must be made more effective. Knowledge acquisition must be embedded into, and integrated with, knowledge utilization in order that learning be functional and relevant.

---

*TEMP = Technology, Engineering, Mathematical sciences (including computer science), Physical science.

(II) New, metacognitive educational considerations must be explored and given greater prominence in order that students be able to parlay their current mathematics education and beginning career status into productive future learning and professional growth.

(III) Changes in perception, attitudes, and role models are needed in order to realize (I) and (II).

These beliefs reflect and interface with important aspects of classroom instruction, artificial intelligence research, and cognitive (including neurobiological) research. Unfortunately these nodes of mathematical endeavor are not as well interrelated as they could be. Thus, it is my hope that this conference, and this paper in particular, will help to stimulate further interest in strengthening these connections.

In this paper, I attempt to (loosely!) depict mathematical knowledge as the resultant vector whose components are interactive processes such as the acquisition, representation, utilization, organization, and management of information. For each person, the coordinates of these component vectors are individual-matrix dependent. Accordingly, mathematical knowledge should be thought of as a dynamic vector that grows and changes orientation in one's intellectual space.

The instructional strategies advocated in this paper are intimately intertwined with behavioral objectives, information-processing, and styles of learning. They are offered as general principles that can enhance mathematics instruction for all students. For TEMPs, these approaches should be viewed as first-order guiding principles that constitute the logical prerequisities and pragmatic basis for higher order considerations - including, for example, metacognition and learning how to learn - that will be the focus of a forthcoming paper in progress.

New Awareness

The present crisis in college mathematics instruction is not so much one of "what specific course content and to whom it should be taught" as it is a reflection of continued failure by the mathematical community to properly communicate what mathematics is and how it can be of value to different, changing student populations.[1] This is sine qua non! Without such understanding and guidance, students will find easier or more rewarding academic disciplines beckoning; why bother with mathematics and its demands?

Most high school students and college freshmen are curriculum captives insofar as they must usually complete certain mathematical course requirements. But given their first opportunity to make choices, college sophomores and juniors increasingly vote to abandon mathematics by enrolling in other courses of study [32]. As adults, they'll also vote with their political and financial influence. These votes have ominous implications for the future concerns and allocation of resources for college mathematics education.

The first two years of college mathematics is particularly crucial for influencing and partially reversing these voting patterns. However, new perspectives and attitudes are required to bring about such changes. Indeed, it is my belief that the

---

1. "...the number of Native Americans, Hispanics, Orientals, Mormons, and Seventh Day Adventists are all increasing rapidly." "...any surge of new enrollments during the next two decades in higher education will be led by minorities, particularly blacks and Hispanics." "In most community colleges today, the average age of students is thirty-six and climbing." For further details and a demographic portrait of students in the 1990s, see [8].

two most critical factors in teaching mathematics concern "what" one conveys and "how" communication takes place. Both factors are intimately intertwined with information-processing and learning; each has affective as well as cognitive dimensions.

> "What" one communicates in mathematics instruction transcends the elucidation of mathematical concepts; the teacher of mathematics also conveys (consciously and unconsciously) a great deal to students about the intrinsic nature and value of the discipline itself. Students' impressions and attitudes about mathematics play an important role in their motivation (therefore, commitment and perseverence) and ultimate success or failure in mathematics courses. (G

Thus, effective mathematics instruction must begin by making students want to study mathematics.

> "How" one communicates in mathematics instruction goes beyond the exchange of ideas and information. Classroom learning experiences and attitudes give rise to long-lasting psycholosocial values on what it means to do mathematics and who should do it. (G

Implicit in each of these factors is the realization that effective learning is rarely possible if teachers of mathematics cannot introduce and develop concepts in a manner commensurate with their students' information- processing abilities and levels of understanding. This realization subsumes an awareness of the fact that a large constellation of behavioral patterns may be at work in predisposing students to success or failure in their mathematics courses, particularly so at the basic skills level.[2]

---

2. In "Mindscape and Science Theories" [11], Maruyama uses the term mindscape to mean "a structure of reasoning, cognition, perception, conceptualization, design, planning, and decision making that may vary from one individual, profession, culture, or social group to another." He distinguishes four pure mindscapes and their combinations, and illustrates their aspects at the overt, covert, and abstract levels.

O. J. Harvey administered psychological tests to university students over a number of years. In [7], he identified four epistemological types and their distribution among first-year university students.

Thus:

Teachers of mathematics must appreciate individual differences and understand how psycho-physio-social factors impact on styles of learning. $(GP)_3$

In this vein, it is singularly important for instructors to realize that they too have their own cognitive preferences. The types of exams they prefer and develop, for instance, reflect their own cognitive styles and not necessarily those of their students. Thus, students' success (or failure) may not depend only on course content, but may also be related to the information-presenting strategies and instructional demands of their teacher. People do learn to learn differently! Instructional procedures which may be beneficial to some students can disadvantage and be counterproductive to other groups of students. Behavioral differences <u>must</u> be taken into consideration, if people having different styles of learning are to interact fruitfully.[3] A few simple examples suffice to illustrate this point.

The quality and quantity of interaction in the classroom are important ingredients for learning. While some students prefer, and do better, working alone, others learn best through some form of give-and-take. The nature of interaction conducive to learning will vary according to the student's background and psychological profile. Since setting, ambience, and interaction are interrelated, it is not immediately clear if students'classroom inactivity result from culturally-related

---

3. An appropriate modification E. T. Hall's statement [6] is: most instructors are only dimly aware of the elaborate and varied behavioral patterns which prescribe our handling of time, spatial relations, and our attitudes toward work, play, and learning. Accordingly, we insist that everyone else do things our way ... and those who do not are often regarded as "underachievers."

reasons, because they are consciously (or unconsciously!) separating and disinvesting themselves from classroom instruction, or because they feel anxious and uncomfortable in the educational environment.[4] Instructors alert to these nuances can enhance learning through classroom-teaching strategies that are appropriate to their students' behavioral needs. (For examples of student-student and teacher-student interactive strategies, see [13], [14], [28].)

Unfortunately, most instructors require that all students take the class exam at the same time, despite the fact that individuals learn and grow at different intellectual rates. This requirement clearly stacks the odds against the slower learners as well as those who (appreciating time other than as a preciously dwindling commodity)[5] have not yet learned to plan sufficient time for study. Exam grades for these "out of phase" students do not reflect their actual subject mastery once such students have caught up. Accordingly, their final grade - based on grades which reflect their states of unpreparedness - may not be commensurate with their knowledge at the end of the course. This disadvantage can be diminished, if not eliminated, by

---

4. Styles of participation conducive to learning also vary with culture. North American Indians learn best through observation. Oriental students seem to do well without heavy emphasis on classroom participation. Americans generally require more interaction than students from Anglo-French cultures, but not as much as Hispanic students.

5. Variations in the perception and utilization of time become evident as one moves westward and southward from the northeastern part of the United States. As a rule, however, Americans think of time as being linear, sequential, and quantifiable. "It should take x time to cover this material; we'll plan an exam for y date." Other cultures share neither our sense of urgency nor our immutable compartmentalization of time: it makes more sense to disregard time constraints and work at the job until it is completed than it does to abandon one unfinished task in order to begin a new one.

broadening the constraint of "fixed day" for an exam to "fixed period" for that exam. For example, students can be given the opportunity to take one of three variants of the exam (test T_i on day D_i for i=1,2,3) during a fixed exam week. In this humanistic context, exams can do more than attempt to subjectively quantify levels of understanding; they can (and should!) be used as pedagogical tools for motivating and rewarding further learning. For instance, students who did poorly on an exam will be highly motivated to clear up specific areas of weakness if they are allowed to take another variant of this exam (during its fixed test period, for example)-in which case, their overall grade for that class's exam is the average of the two exams taken. For another variation on this theme [15], students can be made aware of the fact that each class test will contain one arbitrarily chosen problem from each of the preceding class tests.

Knowledge Transmission and Acquisition

The notion of learning has a wide range of interpretations among people - both in terms of what "knowledge" means and what is required to reach that state of knowing. Unfortunately, far too many students view mathematics as a lifeless body of facts and formulas to be memorized or stored for short-term, cued recall; doing mathematics is too widely interpreted as concept-identification, formula substitution, symbol manipulation, and problem solving in a very narrow, artificial doman. Why is this so? Why have so many students been lulled into these misconceptions, and how can we help them to better appreciate what mathematical knowledge means and what is required to reach that state of knowing?

Each of the above questions must have a multiplicity of answers. But surely what instructors expect and demand of students is pivotal. Thus, we must accept the responsibility for this imprinting and we must take the initiative for bringing about some very fundamental changes in our students' perceptions. A necessary first step is to make it convincingly clear that:

Knowledge acquisition does not imply knowledge utilization.

Just being able to identify a geometric figure (say, a rhombus) reveals nothing about the intrinsic properties of that figure. And symbolic manipulation without understanding is only slightly more meaningless than solving a trivial variant of the same problem for the twentieth time. That such superficial forms of knowledge are minimally functional can easily be demonstrated, and must be driven home, by instructors. It is also very important to alert students to the impact of a powerful anxiety-reducing drug, commonly called 'pocket-calculator.' It alleviates students' motivation to learn by making them feel that they can use it to solve all their mathematics problems. This myth is also easily dispelled. For example:

Problem 1. Enter any number $x > 0$ on your calculator and repeatedly use the $\sqrt{\ }$ key. What do you get? Why?

Problem 2. On your calculator, enter 2 and take $\sqrt{\ }$. Continue to repeat this pattern of adding 2 followed by taking $\sqrt{\ }$. What do you get? Why?

Problems which can be posed, but not solved, by a calculator are effective for demonstrating to students that their head-held calculator is much more powerful than their hand-held calculator and that although calculators can be helpful for computing, they should not be antidotes for the headache of having to think.

Dispelling students' myths is not enough. There still remains the question of how to help them appreciate mathematics as a dynamic and multilayered activity - a richly rewarding and evolving synergism of process and product. This we now consider in greater detail.

In most instances, mathematics instruction is considerably more effective when several modes of perception are used - as may be the case, for example, when (left-hemispherically oriented) technology students and, say, (more right-hemispheric) humanities majors are in the same course.[6,7] Thus, both the symbolic-analytic approach and the visospatial-relational approach may be used to prove (Figure 1) that "the geometric series $1 + \frac{1}{2} + \frac{1}{4} + \cdots$ converges to 2."

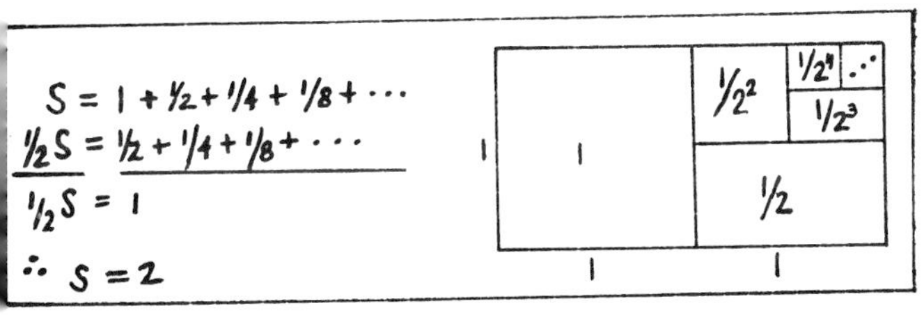

Figure 1.

If $\sum_{n=1}^{\infty} 1/n = S$, then

$S = 1 + 1/2 + 1/3 + 1/4 + \cdots$

$> 1 + \frac{1}{2} + \cdots = S$

yields the contradiction $S > S$.

Figure 2.

In the same spirit, analytic proofs (viz. converging-series tests) that the harmonic series diverges may be supplemented with, or made more plausible by, following a (right-hemispheric) analogical tact as in Figure 2.

---

6. Today, it is well known that there exists major differentiations of functions between the brain's left and right hemispheres. In the most simplistic terms, left-hemispheric thinking resembles the discrete, sequential processing of a digital calculator; right hemispheric thinking simulates the concurrent, relational activity of an analog computer.

7. Cohen [2], [3] found that white middle-class children tend to be analytical in orientation, whereas Chicano and black children tend to be rational. She also found difference in orientation among professions [4].

We all know that $\sum_{k=1}^{n} k = n(n+1)/2$ can be (and usually is) proved by induction. But, as is often the case, students feel cheated: "here, induction is an accessory after the fact. How did one know the formula to be verified in the first place?" Instructors, of course, can invoke Gauss' (more right-hemispheric) relational approach to obtain the aforementioned conjectured formula for verification (Figure 3).

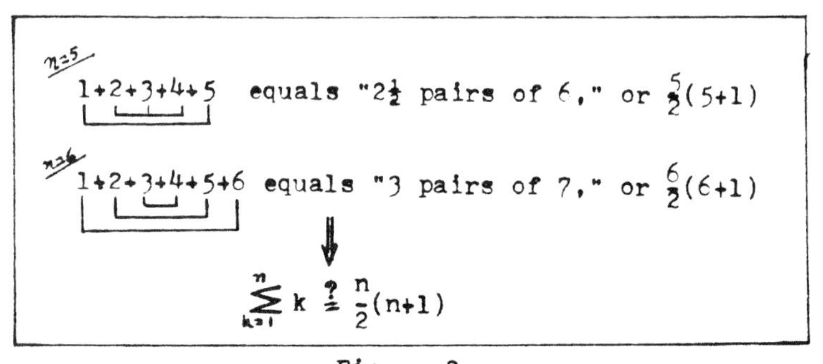

Figure 3.

One can also obtain $\sum_{k=1}^{n} k = n(n+1)/2$ by counting the dots in the right triangle of Figure 4. (The right triangle of dots, when reflected with respect to its hypotenuse, produces a square of dots plus an extra superimposed diagonal. Thus, $2\sum_{k=1}^{n} k = n^2+n$.) Figure 4 also illustrates some of the author's visually-induced proofs of other known results [21].

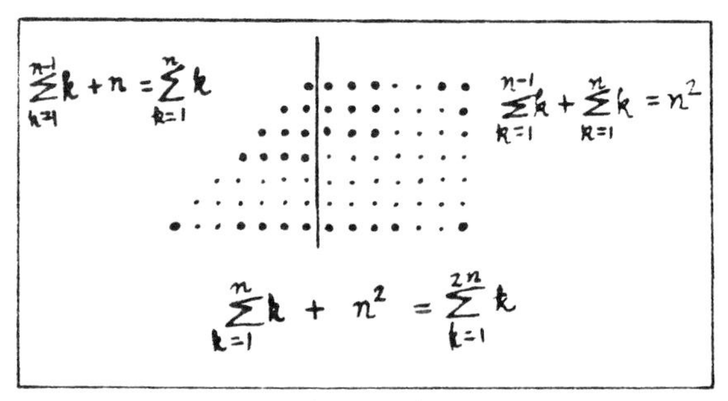

Figure 4.

Since pictures are usually more easily recalled than rocedures, visual portrayals of algebraic processes can enhance he retrieval of information. Synthetic division and synthetic ultiplication [10] offer two good illustrations. The point eing emphasized is:

> Multimodal representation of concepts can do more than convince students of concepts' veracity; they have the potential for synergistic learning - as, for example, when concepts are introduced by one modality and students are asked to finds representaions/proofs in other modes of thought.

(GP)$_4$

But let's not stop here! This leads in a natural manner to a hole new dimension of thinking. An example or two suffices to ake this clear.

1. Having symbolically demonstrated that $S = 2$ (Figure 1), he author was surprised to see that some students felt tricked nd less than convinced of this result. "How did you know to ultiply by 1/2 and then subtract?" "Where did the whole series isappear to?" Interestingly, the visual proof- stumbled upon uring class session - was perfectly acceptable to everyone. A ew additional remarks, between pauses, began to lead students to new awareness. It soon became clear that our visual proof was lso "an accessory after the fact." How, after all, did I know o begin with a 1 x 2 - sized rectangle in the first place? The ymbolic proof was also challenged as being bogus since it too as based on the aprioi knowledge that S was a finite number. ext, we also discovered that the same algorithmic process can roduce meaningful as well as meaningless results (replacing the atio "1/2" by "r" for r>1, we still obtain $S = 1/(1-r)$), and that lgorithmic, existential, and constructivistic thinking are ntimately interrelated. Finally, it was intuitively clear that he analytic proof generalizes much more efficiently than a

geometric one to arbitrary converging geometric series (Students may enjoy attempting a visospatial proof, or they ca refer to the author's discovered generalization [17].)

2. Given the motivation and opportunity to experiment, eve the weakest students will quickly discover that the distributiv multiplication depicted below

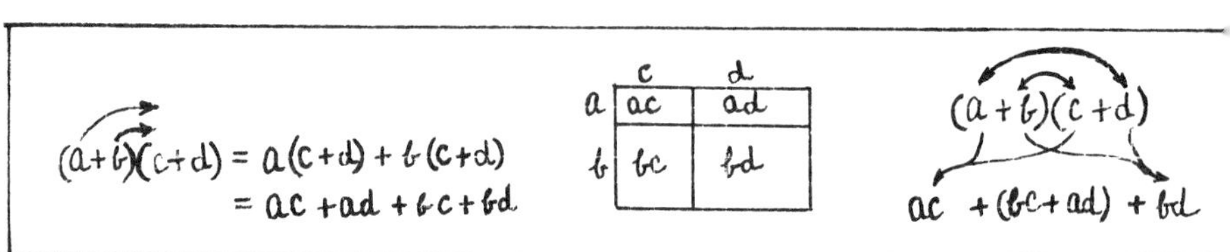

Figure 5.

is the representation that can best be extended to the mult lication of multinomials.

3. A few well-selected examples made it clear that Gauss combinatoric approach (of using pairs of numbers) had greate potential for adaptation to other contexts than do dot proofs but it was not as pervasive as mathematical induction. (Here wa the beginning of a new appreciation and respect for induction.)

Comparing and analyzing the efficiency, extendability, and generalizability of representations is an important first step toward developing the types of awareness students will need in their algorithmic and computer-related mathematics learning. (See, for example, [9].)

Experimenting with alternate modes of representation can also be stimulating and informative to instructors. Figure 4, for example, yielded a newly discovered visual proof by the author of the fact that *the sum of the first n terms of an arithmetic progression plus n^2 times the difference equals the sum of the next n terms of the progression.*[8] By examining the various representations students use, we can better judge how well they understood the concept in question.[9]

Embedding concepts in processes can help students appreciate mathematics as a dynamic and multilayered activity - an evolving synergism of process and product. These perceptions must begin, so to speak, at the molecular level. Numbers, variables, shapes, formulas and equations, as well as, other such basic entities, must not be perceived as passive, static notions, but rather as interactive processes and actions. This impacts on how information itself is presented.[10] To use Herb Simon's analogy [29].

A physician's knowledge of how to treat diseases is useless if the physician can't tell when the patient has the disease. Thus, a large part of medical knowledge consists of condition-action pairs; the condition being the disease symptoms, and the action being the appropriate treatment.

---

8. If the dots on the hypotenuse of the right triangle are labeled "a" and all other dots on the trapezoid are labeled "d" then (since the dots on the triangle plus the dots on the square comprise the dots on the trapezoid) $\sum_{k=0}^{n-1}(a+kd)+n^2d=\sum_{k=n}^{2n-1}(a+kd)$.

9. Greeno [5] offers three general criteria for judging the degree of understanding of a represented concept: internal coherence of the representation, its connectedness to other relevant knowledge, and how accurately it captures the concept's essential features.

10. Too many students think of a variable x as being a fixed unknown (rather than as an actively roaming entity - an operator whose character changes depending on where it is encountered in its domain); formulas are perceived as receptors passively waiting for substituted numbers (rather than as the algebraic or visually portrayed embodiments of how variables relate to each other); and equations are considered as fixed states of equilibria (rather than as reversable processes, where each side eyeballs the other and can get there by an appropriate sequence of transformations).

This is not the format of mathematics/science knowledge, in general. We are much more explicit in enunciating principles than in describing when and how they can be applied. Formulas and theorems, for example, do not always carry internal information about contexts or situations that should evoke their use. Greeno [5], is probably correct in his impression that "most teaching of algorithmic processes often focus almost entirely on the actions to be performed, with little attention to the issue of when to perform them." Mathematics texts, on the other hand, seem to assume that once students are shown a few worked out problems, they'll be able to generate their own situation-action responses for solving problems. This is not always the case, and even less so for students in their earlier college mathematics courses.[11]

The point being stressed here is that every important mathematical result should be presented as the action component of condition-action pairs. For such a "production," the conditions needed for the result to apply are built into the presentation. In broader terms:

---

11. For instance, knowing that $x^0 = 1$ is useless in Problem 1 if students don't realize that $x^{1/2^n} \longrightarrow x^0$. Knowing how to solve quadratic equations is useless in Problem 2 if students don't realize that $y = \sqrt{2 + \cdots \sqrt{2 + \sqrt{2 + \sqrt{2}}}}$ can be expressed as $y^2 = 2 + y$. Finally, students' knowledge of the Pythagorean theorem is useless they cannot use it in appropriate situations (Problem 3) and they attempt to apply it to inappropriate contexts (Problem 4).

<u>Problem 3</u>. Using only a compass, measure off length $\sqrt{3}$ along AB.

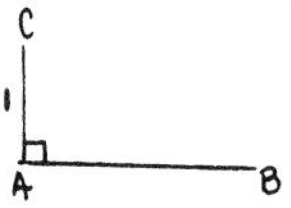

C

1

A B

[Solution: Mark off D on AB such that AD = 1; then CD = $\sqrt{2}$. Mark off E on AB such that AE = $\sqrt{2}$; then CE = $\sqrt{3}$.]

Every key notion and every important principle should not only be considered in terms of its intrinsic properties, but also as the basis for solving a primitive class of problems. $(GP)_5$

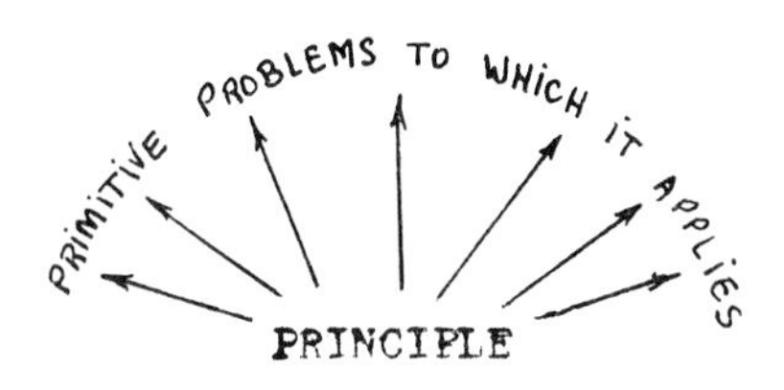

Figure 6.

A nice illustration of this principle is "An Approach to Problem-Solving Using Equivalence Classes Modulo n" by J.E. Schultz and W.F. Burger [29].

There is another important pedagogical facet to $(GP)_4$ – namely, the manner in which this type of thinking and awareness can be broadened to solve problems. Indeed, it is well known that the manner in which a problem is described is of critical importance in determining how easily the problem can be solved or whether it can be solved at all.

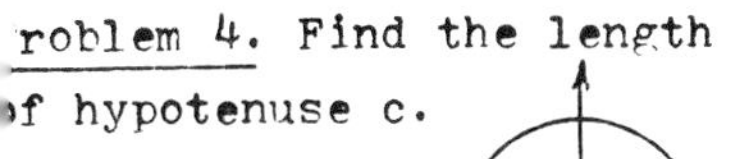
roblem 4. Find the length of hypotenuse c.

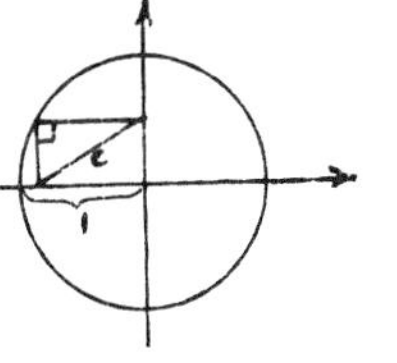

Figure 7.

Problem 5. Find the area of the parallelogram plus the area of the square.

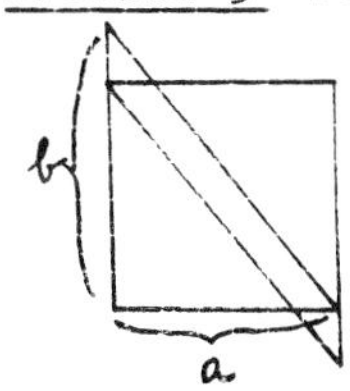

b

a

Figure 8.

Attempts to solve Problems 4, 5 corroborate the findings of research experiments: subjects don't ordinarily search for the most efficient representation of the problem; they tend to adopt the representation of the problem from the language of its statement.[12] Thus, as Simon points out [30], it should be clear that:

> Instructors need to help students improve their skills in reformulating and restructuring problem representations. It is most important to make students understand that the value of their mathematically-related career skills will, in large part, depend on their ability to recognize and construct contexts that evoke appropriate mathematical principles and processes.

As the instructional dual to $(GP)_5$, where principles served as "seekers" of conditions and contexts where they apply, problems can serve as "attractors" for as many distinctly different solutions as possible.

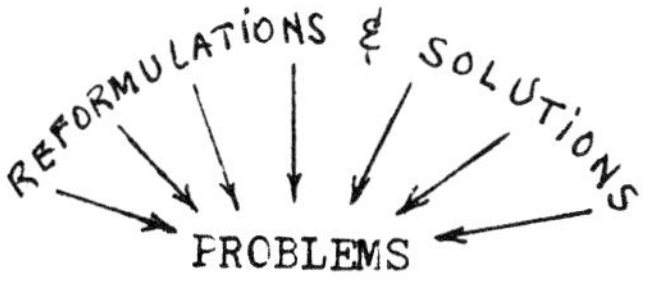

REFORMULATIONS & SOLUTIONS

PROBLEMS

Figure 9.

An especially nice illustration of this is J. Staib's "Answer Finding Versus Problem Solving" [32], where the class discovered nine different ways to find the distance from a point to a line. Also see "Convexity in Elementary Calculus: Some Geometric Equivalences" [1], and Pedersen and Pólya's "On Problems with Solutions Attainable in More Than One Way" [25].

---

12. In Problem 4, the "hypotenuse" c equals 1 since the other diagonal of the rectangle is the radius of the circle. If the "parallelogram" and the "square" (in Werthheimer's Problem 5) was restructured as overlapping right triangles of base a and height b, the desired area is immediately seen to be ab.

The perception of objects, systems, and processes vary considerably amongst people, and this has tremendous bearing on how mathematical notions are perceived and utilized. This is especially true in the classroom. As teachers are expounding on mathematical notions and principles, students are busily concocting their own idiosyncratic versions based on their own consistent private logic. Such cognitive misinterpretations, however, are not confined only to developmental mathematics students or to those whose backgrounds do not reward clear and precise thinking. In the margins of Bourbaki's advanced level texts, for instance, the roadway danger signal ☡ (caution!) is followed by elaborative comments designed to help prevent readers from making wrong interpretations that are consistent with the antecedent exposition. The point being stressed here is that:

> In presenting information, it is vitally important for teachers to anticipate and preempt students' misinterpretations. *(GP)*$_7$

The instructional strategies summarized in (GP)$_4$ - (GP)$_7$ can help teachers monitor, and become more attuned to, the nature of these misinterpretations. To the extent that we examine, analyze, and modify our instructional strategies, we gain a higher form of instructional knowledge and an increased capacity for becoming better <u>imparters</u> of knowledge.

There are also important information management considerations for the <u>acquirerers</u> of knowledge. Consider, for instance, students who do well on homework assignments or quizzes covering each specific aspect of a problem situation but still do poorly on exams where they are no cues as to which solution strategies to apply to the problems as a whole. In short, they lack certain aspects of control knowledge (that is, information management). Other manifestions of deficiences or weaknesses in control knowledge include: incorrect or incomplete categorization of problem protoypes, lack of coherent knowledge structure and organization, inability to recall or retrieve information, nonassessment of concept attainment, and disregard for solution verification.

There are many strategies for helping students to overcome these deficiences. A contextually-representative sample might be the following:

- Classroom exams. " Without actually solving problem P, carefully describe and/or set up in as many different ways as possible how to obtain the answer to P."
- Homework assignments. "Compare and contrast the types of problems (and how they are solved!) in this chapter with those in . . ."
- Interactive discussions. "How do you know that your method is correct? Your answer is reasonable? ..." [13]
- Term papers or course-related projects. "Summarize the chapter's (course's) key concepts and principles, and be sure to discuss or depict their interrelationships." (See, for example, [26], [27].)
- Realistic role models "...Okay, I'll try to solve and analyze this mathematical problem you've encountered in the physics lab (on the job, for contest X, ...). I'm not really sure where to begin...Suppose we first attempt...because ... "

---

13. In "Questions in the Round - An Effective Barometer of Understanding" [14], the instructor proceeds around the room requiring that each student either ask a question (to be answered by the instructor) or else be asked a question by the instructor. This strategy provides an excellent opportunity for instructors to ask questions of control knowledge.

The most effective strategy, however, is to make students alize that while it is natural to form misconceptions and make rors, specific actions for their detection and analysis are so important mathematical activities. We must demand, and udents must be made to appreciate, that verification and alysis are necessary in doing mathematics. Thus:

Control knowledge must be appreciated as being an integral part of knowledge acquisition and accumulation. $(GP)_8$

It seems clear that both teachers and students can receive and part important types of knowledge from each other. Accordingly:

Teachers must invite and encourage students to be responsible partners in an interactive collaborative learning environment. $(GP)_9$

Interactive and collaborative aspects of $(GP)_9$ have already en considered earlier. The invitation I urge is not explicit nature, but rather implicit in the way we teach and do thematics in the classroom - manifestations,soto speak, of ing "great teachers" in the sense of J. Epstein's edited volume essays Masters: Portraits of Great Teachers [12]:

"What all the great teachers appear to have in common is a love of their subject, an obvious satisfaction in arousing this love in their students, and an ability to convince them that what they are being taught is deadly serious."

The most natural embodiment of $(GP)_9$ is for teachers to guide, sist, and/or collaborate with students in actually doing thematics that has meaning to them. There are many ways to oceed, depending on the students' capabilities and levels of thematical sophistication:

Mathematical problems, puzzles, and games have been popular nce antiquity, and their solutions have contributed much to the velopment of modern mathematics. Thus, Leibnitz appears to have been correct when he said, "Men are never so ingenius as when they are inventing games." Recreational mathematics and examples from everyday life always stimulate students' curiosity and whet their intellectual appetites for more.

Weaving mathematical tapestries can be fascinating. Combining and interlacing novel ideas from diverse areas of mathematics (as distinct from applying mathematics to other disciplines) is a beautiful way to impress students with the fact that mathematics is indeed a coherent, harmonious whole.

Doing mathematical research cannot fail to convey the challenge and excitement of attempted discovery. Fruitful research exists at all levels.[14] The rewards of successful research - giving an invited (classroom) lecture, seeing one's results(s) in publication, and other forms of peer acknowledgement - can be the biggest payoffs and reinforcers for students to stay invested in the study of mathematics.

Concluding Remarks

Finally, as we began, let us pause to reflect on where college mathematics could be heading. To the extent that we succeed in going beyond changing our students' votes and actually imbue our more capable students with positive perceptions of (and feelings toward) mathematics, we increase the likelihood that the focus of mathematics instruction will not only be as a "seeker" of contexts and domains of application, but will also

---

14. For examples of mathematical research that can be undertaken by, or shared with, students in their earlier years of college mathematics, see ([18], geometry), ([21], precalculus), ([19], calculus), ([20], number theory), ([23], statistics), ([22] and [24], general).

become an "attractor" for significant contributions from many of these serviced disciplines. Mathematically competent and well predisposed students entering careers in computer science, the social and biological sciences, and the humanities will most likely be more motivated and better equipped to bring their expertise to bear on improving and enhancing mathematics instruction.

By giving careful attention to the _what_ and _how_ factors of mathematics education, college mathematics instructors can play an important role in the evolving vitality and future growth of mathematics instruction at all levels. It is not an opportunity that should be cavalierly disregarded.

References

[1] Victor A. Belfi, "Convexity in Elementary Calculus: Some Geometrical Equivalences." _The College Mathematics Journal_ 15:1 (January 1984) 37-41.

[2] R. A. Cohen, "Conceptual Styles, Cultural Conflict, and Nonverbal Texts of Intelligence." _American Anthropologist_ 71 (1969) 825-856.

[3] ________________, "The Influence of Conceptual Rule-sets on Measures of Learning Ability." In _Race and Intelligence: Anthropological Studies_ 8 (1971) 41-57.

[4] ________________, "Relational and Analytic Intelligence Among Designers and Researchers." In _Environmental Design Research_ Vol. 1 _Selected Papers_, edited by Wolfgang Preiser; Dowden, Hutchinson and Ross Publ. Stroudsburg, PA. 1973.

[5] James Greeno, "Understanding and Procedural Knowledge in Mathematics Instruction." _Educational Psychologist_ 12:3 (1978) 262-283.

[6] E. T. Hall, _The Silent Language_. Anchor, Garden City, NY, 1973.

[7] O. J. Harvey, _Experience, Structure and Adaptability._ Springer, New York, 1966.

[8] Harold L. Hodgkinson, "Guess Who's Coming to College." _AAUP Academe_ (March-April 1983) 13-20.

[9] Dan Kalman and Warren Page, "Efficient Exponentiation Algorithms for Calculators." _The College Mathematics Journal_ 16:1 (January 1985) 58-60.

[10] Kenneth R. Kundert, "Why Not Teach Synthetic Multiplication?" The Two-Year College Mathematics Journal 11:2 (1980) 121-122.

[11] Magoroh Maruyama, "Mindscapes and Science Theories." Current Anthropology 21:5 (October 1980) 589-600.

[12] Masters: Portraits of Great Teachers, ed. Joseph Epstein, Basic Books, New York, 1981.

[13] Warren Page, "Small Group Strategy for Enhanced Learning." American Mathematical Monthly 86:10 (1979) 856-858.

[14] ______________, "Questions in the Round: An Effective Barometer of Understanding." The Two-Year College Mathematics Journal 10:4 (1979) 278-279.

[15] ____________, "Exams Can Leverage Learning." The Two-Year College Mathematics Journal 10:1 (1979) 38.

[16] ____________, "Count the Dots." Mathematics Magazine 55:5 (March 1981) 97.

[17] ____________, "Convergent Geometric Series," Mathematics Magazine 54:4 (September 1981) 201.

[18] ____________, "Compass Construction $\sqrt{n}$." The Mathematics Association of Two-Year Colleges Journal 10:13 (1976) 180-181.

[19] ___________, "The Formula for Arc Length Does Measure Arc Length." In Two-Year College Mathematics Readings, ed. Warren Page, The Mathematical Association of America, 1981, pp. 111-114.

[20] _____________, "A General Approach to $p \circ q$ r-Cycles." In Two-Year College Mathematics Readings, ed. Warren Page, The Mathematical Association of America, 1981, pp. 263-274.

[21] Warren Page and Leo Chosid, "Synthetic Division Shortened." The Two-Year College Mathematics Journal 12:5 (November 1981) 334-336.

[22] Warren Page and Harold Dorwart, "Numerical Patterns and Geometrical Configurations." Mathematics Magazine 57 (March 1984) 82-92.

[23] Warren Page and Vedula N. Murty, "Nearness Relations Among Measures of Central Tendency and Dispersion." The Two-Year College Mathematics Journal: Part I in 13:5 (November 1982) 315-327, and Part II in 14:1 (January 1983) 8-17.

[24] Warren Page and K.R.S. Sastry, "Polygonal Numbers on a Square Lattice." To appear.

[25] Jean Pederson and George Pólya, "On Problems with Solutions Attainable in More than One Way." The College Mathematics Journal 15:3 (June 1984) 1218-228.

[26] Frederick Reif, "Teaching Problem Solving-A Scientific Approach." The Physics Teacher (May 1981) 310-316.

[27] Frederick Reif and Joan I. Heller, "Knowledge Structure and Problem Solving in Physics." Educational Psychologist 17:2 (1982) 102-107.

[28] Alban J. Roques, "Homework - A Problem With a solution." The Two-Year College Mathematics Journal 10:2 (1979) 116.

[29] James E. Schultz and William F. Burger, "An Approach to Problem Solving Using Equivalence Classes Modulo n." The College Mathematics Journal. 15 (November 1984) 401-405.

[30] Herbert A. Simon, "Problem Solving and Education." In Problem Solving and Education: Issues in Teaching and Research, eds. D. T. Tuma and F. Reif. Lawrence Earlbaum Associates, Hillsdale, NJ, 1980, pp. 81-96.

[31] John Staib, "Problem Solving Versus Answer Finding." In Two-Year College Readings, ed. Warren Page, The Mathematical Association of America, 1981, pp. 221 227.

[32] Gail S. Young, "Who Takes Elementary Mathematics Courses? Why? A Guess and Some Problems for Change." In The Future of College Mathematics, ed. A. Ralston and G. S. Young, Springer-Verlag, New York, 1983, pp. 13-26.

(DISCUSSION BEGINS ON P. 367.)

DISCUSSION

THE LEARNING OF MATHEMATICS

1. Learning to Learn
2. "Green Globs"
3. Cognitive Science for Teachers
4. Preconceptions
5. Cooperative Learning

LEARNING TO LEARN

Kaput: There's a very close relationship between cognitive science (and even A.I.) and computer science, of course, and the style of research is more attuned to the individual. We are now able to do detailed, intense observations of individuals building models, formal and informal, that represent the hypotheses going on in their heads when they are doing mathematical tasks. And that's very different in tone and impact, and more important for us all than what went on earlier. Unfortunately it's only a few years old, and is not yet bearing a lot of fruit; the main fruit that it has borne so far has been negative. When we started to look closely at what students really know and are able to do, we discovered a mess. The students don't really understand variables. They don't understand variation. They don't understand much of what we think we are teaching them! In fact, we confirm our worst fears--basically students are learning to push symbols.

"Green Globs"

"Green Globs" [algebra software] is an example of how technology can help solve a fairly specific pedagogical problem. We've all had the experience of trying to teach the quadratic formula; and we have taught graphing of polynomials, in particular, the graphing of quadratic functions. Somehow the information about the parabolas that students graph resides in one part of the student's mind, and the algebraic solution to equations resides in another part of that student's mind. It's extremely difficult as a teaching task to bring those together. When working with

"Green Globs," the student is given a bunch of green globs on the screen and is asked to type in an algebraic equation whose graph hits as many of them as possible. Without going into the details of it, what it does is compact into an extremely small amount of time an enormous amount of graphing experience while laying off onto the machine the computations that underlie the graphing of a particular point. Now that cannot be the only graphing experience students have, but one of the effects that I have seen in students doing that is that they get really good at adjusting the coefficients of the polynomials so that it has to wiggle at the right height and so on, and for them there is no difference between the equation and the graph. They are both just different aspects of the same thing. It fuses those things in their heads so that they are a single "thing" rather than two things that need to be tied together. I think that it's a very important breakthrough.

Watkins: I have a question for Jim. When you walk into a classroom and you see something like "Green Globs," what you immediately see is a bunch of boys gathered around the screen. There is no doubt about the fact that they love it, and there is no doubt about the fact that it's good math. But my question is: How do you teach the same sort of mathematics to people who are "people people" rather than "computer people"?

Kaput: I think that part of the problem right now has to do with the fact that the "Green Globs" stuff is not very heavily distributed and so there are very few opportunities to use it.

If they're everywhere and there is a wide variety of software opportunities, not just "Green Globs," then I don't think you'll have that problem of dealing only with a small group of people.

Watkins: Don't you believe that there are sex differences?

Kaput: I don't think so. If every student gets the opportunity to do it, and it's in the classroom with everybody, and it's a natural part of everybody's life, it will happen.

Page: I do not think that you can just take A.I. and the latest studies in artificial intelligence-related research, and say: "Well, this is going to apply to individuals." A fundamental belief for myself is that artificial intelligence can prescribe new types and new perspectives of research which are very useful, but I think that A.I. needs the cognitive research, the way people think, the way they feel, in order to make it rooted in the realism of the system which is a biological system--the human being. So, in other words, to make A.I. work, you need the cognitive research as to how people work, how people think, what are sex differences, what is the difference in brain lateralization, what is cultural imprinting, etc.

Cognitive Science for Teachers

Maurer: Jim, I want to take you up on a sentence of your paper, where you were talking about what will have to go out and what will have to go into the curriculum. "It is imperative to know how these different subjects relate to one another in the minds of the student." More generally, do you feel it is imperative for us as educators to know about what is discovered by cognitive science research? Let me argue that that's nonsense; that it's neither imperative nor necessary, and merely valuable. Now let me give two arguments for that. For several thousands of years of civilization, without knowing much about how people learn things, most people seem to learn most of what they need to know and one can even make a biological argument that it's due to evolution. We would have died out had we not somehow had the capacity to learn despite lack of real knowledge about how to educate. Secondly, one can argue that, even if it is the case, once we know this additional knowledge it may be a small variable compared to variables like--Does the teacher care?--How large is the classroom?--Is the teacher feeling professional? etc. I think that the second argument counts, and I think that the first argument counts too, because despite what you said it is not true that it is just in the last ten years that there is much more information than anyone can ever learn. That's been true for more than a hundred years.

Kaput: We haven't tried to teach it to very many people, however. It's been a very narrow segment of the population that we've been trying to teach until relatively recently.

Maurer: It's a very small fraction of the population to whom we've been trying to teach a large amount of technical material, until fairly recent times. But it's not clear even in the future how large a segment of the population is going to need or going to have a high technical base.

If this is true--that it is not imperative to know all this--but that it's merely valuable, then I'm not sure what to make of your earlier comment about professionalism. You said that the way teachers could be more professional is for them to be involved in the science of learning. The best I can make of that statement was that you wanted all teachers, in some sense, to be second-level cognitive scientists, and I am not sure that this is even possible or actually valuable.

Preconceptions

Kaput: I think there's a good argument here as to whether or not it's absolutely imperative or just extremely valuable to know how we can map knowledge in the student's head. One point I did make today, which is in my paper and is one of the fundamental consensus points of recent research, is that students come into our classrooms with a fairly complete and extremely powerful, robust, and resilient collection of preconceptions by which they interpret anything we say or do in the classroom. Knowing the construction and the organization of that collection of preconceptions is critical to being able to deal with it in a systematic way. I don't think that you would want to take your laissez faire attitude towards the construction of, say, an airplane--particularly the one you're going to fly in tomorrow. You want to be very systematic about that and know everything you possibly can about the metal and how it's organized, etc., before you start even test flying that airplane. We somehow have a laissez faire attitude that is rooted in that fatalism I talked about earlier; a laissez faire attitude about what constitutes student knowledge. I think that we've got be very careful and figure out what's going on in their heads and how that knowledge is organized and structured.

Let me give you a rather simple experiment to indicate that there are radical differences in the way that knowledge is structured in the mind and how it is used. Suppose I gave you the task of listing all the states in the United States: A very verbal person like, say, Steve Maurer would come along and say,

"Alabama, Alaska, Arizona . . ." and so on. Somebody else, a more visual person, might start to say "Maine, New Hampshire, Vermont, Massachusetts, Rhode Island, Connecticut . . ." and so on. There we now have some external information that we can use to make inferences about the structure of the knowlege inside the person's head. For one person, it's a lexical organization of some information based on the organization of the alphabet; the other person is working off a mental image of a map, however sketchy, of the United States--two very different types of structures of information, radically different organizations of knowledge in the head; we haven't begun to sort them and figure them out. If we did, we'd be a heck of a lot better as teachers and curriculum developers. It's particularly important when we get into the business of large-scale software construction--for then we need a very good theory of the mind; we need a very good theory of the subject matter domain; and we need a very good theory of how those two interact.

Maurer: I'd still argue you need a good sense; I'm not sure that you need a good theory.

Tucker: What's coming out here is the fact that every single individual deserves the attention, the probing of their mind and the interest in how they think. I think that there's clearly a trade-off if you do this. Your classroom pace is going to grind to almost a halt. On the other hand, you might be able to slow down the class and take an average student in the middle. There are probably seven or eight categories of ways of organizing information that's common to many of the students. The classroom experience might focus around the student's reactions to problems. One student asks, "How did you think up this problem?" and another student asks, "Why did you think up this problem?" Students can learn by realizing that there are other modalities and ways to think about these things.

Page: How one presents the information--whether it is in small groups, whether it is individual, whether it is interactive, whether it is questioning, whether there are many different representations of the same concept need not take time away from a course. If you can present the idea visually--get another representation--present the information different ways--what you get is a synergism. You get an enriching, you get a whole different perception than when you present something one way, and say "That's the way it is."

Renz: Jim, you gave the example of the use of theory in designing airplanes, but I would submit that, for most airplanes, until very recently, (their) construction was very carefully tested by experiment rather than devised according to some absolute pre-existing theory. And I think the same is true of instruction in the classroom. The new elements that you are talking about are things that we are discovering and becoming aware of, and which, in fact, will guide our practice. It's not as if we had a closed theory that would tell us what to do.

Taylor: One of the things I've observed at the high school level is that remediation programs tend to be pretty much individualized types of programs. We thought that had great promise very early in the game, and frankly we were quite disappointed. I don't know what the research shows at the college level, but the research, particularly at the elementary and junior high levels, is pretty devastating. Shane from the University of Iowa did a kind of meta-analysis and he found that of some 55 subtests on individualized instruction, for about half of them there was no significant difference. Of the remaining ones, by a ratio of about three to one, brand X did better and also the kids that did the least well were the ones who were the least able and the least motivated. They did the least well in terms of just computational skills.

Kaput: One of the problems of individualized instruction is measuring with respect to what parameters. So you don't know what the parameters are until you have an idea of what constitutes, truly, the learning and teaching of that subject matter. People basically let time be a variable and then lose the time flow and the event flow that carries people--some of the slower students along.

Cooperative Learning

Taylor: What I was hearing you talking about, Warren, was something that research recently shows does have an impact; that's cooperative learning, particularly in terms of higher cognitive skills. It also has the structure of a group situation. You can structure the group situation so that you are going to get boys and girls around the "Green Globs" game, and they are going to be working together on it. How do you teach these different kinds of problems of going from a graphical representation to a table, to an equation and so forth? You've got to be able to communicate about it, and one of the things we all discovered as teachers was that we found we learned things *after* we had a chance to teach them. Why? Because we communicate them. So you put kids into a situation where they are communicating. How many of us have seen kids around a computer terminal? The real action isn't between the kid and the terminal. The action is between the kids.

Warren: I'll give you a quick example of teaching methodology in terms of discovery method versus the structured method, which I think relates a lot to how one is going to learn, and also the difference between Steve and Jim as to whether or not over the years we've learned, without knowing, how people learn. Twenty years ago I knew a physics instructor who was having problems teaching the fundamentals of series and parallel circuits. He had a very structured method which was not successful with either the males or the females in his class. One day, instead of doing his usual presentation, he took a brown bag and put X number of 1.5 volt D cells and X number of 1.5 volt bulbs and X number of pieces of wire in there and told the students "Make the lights light." And from that he taught the basics of series and parallel circuits. It was very effective. What he did was match the learning patterns of the students in that class with the problem. He used their learning patterns to establish the knowledge base of series--parallel circuits. I guess that's what I think in terms of Allyn Washington's presentations on teaching mathematics and technical mathematics. It's the application--how you use it--and if you throw it out to the students so that there are multi-modal exchanges within the classroom, you are matching at a higher degree that learning capability of those students to the degree that they can learn.

Gordon: I generally view myself as a technically- or technologically-oriented person; and yet listening to the two of you [Kaput and Page], I find myself reacting negatively on humanistic grounds.

In a sense, I am extrapolating from what you're talking about, and I have little doubts that the level of education can and will be significantly improved, but I am picturing how that will be achieved. The cognitive studies seem to me will lead invariably, particularly with the technological developments, to the point where each student will be uniquely represented as a series of male versus female, Hispanic versus Black versus Amerindian, graphic-minded versus linear, etc. We will be subdividing the students down into individual things; and as you said, I don't want to put words in your mouth, but I'm almost picturing sitting them down for ten hours a day in front of their little boxes, one of them with "green globs," another with "blue globs," another with "red globs"--I am not sure that I like the picture of that society.

Page: That's precisely _not_ the point that I'm trying to make. The point that I am trying to make is that we teach students how to represent information in the way that they prefer, by multimodal representations. For example, if you are a visual person and I present, let's say, the sum of a geometric series, you can learn to think about it geometrically. If Larry is more analytic, that's fine. All I am saying is that there is an immediate consideration here in helping people learn to learn, based on their own individual differences. Obviously, we can't have one specific thing for each individual. But if we learn how to represent knowledge, how to organize knowledge, and how to question it, then when the students have that new technology they're going to do it their own way. That's the key issue.

Fusaro: The mere fact that you are talking about this stuff and think that it is important to study brings a tolerance. How many of you have, for example, looked over the shoulder of students taking notes? We probably want nice neat notes that we all aspire to, and we'd probably correct them if they'd let us. On the other hand, it's been shown without any doubt that that's probably the worst way to take notes. In fact, if you want to forget things you take notes in that manner. Notes should be taken in a graphical fashion as has been studied by an English psychologist. This sounds all very new, but it's really not. Let's have a little fun here and go back 2,500 years to the old Chinese "Ying-Yang" idea which recognized that there is essentially a bimodal approach to everything. There are other breakdowns as well, but the bimodal was to them the critical one. They could have the analytical approach, the graphical approach, the drawing of the map to remember the names of the states or listing them alphabetically. It's a kind of retreat from behaviorism which I welcome. This emphasis on the individual is to be welcomed. As mathematicians we tend to be so damned linear, so damned analytical; I think it's time for us to look at how we approach things.

Rodi: What I'm hearing here is that is not an isolated emphasis on the individual. Not only will it bring tolerance, but if you can have variety within a classroom, it's also going to take students who tend to have one mode of learning and expose them to other modes of learning, and thereby enrich their understanding.

Part 6.

FACULTY RENEWAL AND PROFESSIONALISM

RELEVANCY AND REVITALIZATION: RETAINING QUALITY AND VITALITY IN TWO-YEAR COLLEGE MATHEMATICS FACULTY

by

Ronald M. Davis

SUMMARY

The process of providing quality instruction in mathematics involves more than changing the content of a curriculum as needs change. It must also focus upon ensuring the existence of a committed, confident and motivated faculty with a knowledge of the new or altered content.

The vitality of the faculty is essential for effective implementation of curriculum change. Over the years that vitality has been allowed to decrease through heavy workloads, changing student clientele, increased noninstructional duties for faculty and insufficient remuneration. Two-year colleges must devote themselves to revitalizing the faculty. A quality faculty is the key to quality instruction.

A curriculum change also requires a faculty knowledgeable about the new content. Retraining opportunities, workshops, and minicourses need to be provided with greater frequency and in convenient locations. Opportunities for faculty to experience the activities in local industries need to be provided.

Only a properly prepared and revitalized mathematics faculty will be sufficiently equipped to meet the changing needs of students and industry. This conference must concentrate not only on recommending curriculum change but also on communicating those recommendations and on providing preparation and support for the implementation of these recommendations.

RELEVANCY AND REVITALIZATION

Ronald M. Davis

Retaining Quality and Vitality in Two-Year College Mathematics Faculty

As mathematics instruction in two-year colleges enters the late 1980s and 1990s, two major concerns will come to the forefront. The first of these is retraining faculty so that it will continue to provide instruction relevant to the needs of students in two-year colleges. Retaining this relevancy involves two aspects: (1) the faculty must acquire new knowledge so that it can provide instruction in new areas and (2) the faculty must incorporate new approaches, new methods, or new emphases into existing courses as the need for change arises.

The second concern involves the retaining and uplifting of the vitality of mathematics faculty. Limited resources, changing demographics, and heavy workloads are fast depleting the vitality of faculty in two-year colleges. Revitalizing involves rejuvenating the psychological state of faculty so that they might perform their instructional duties at the best of their abilities.

The Need for Revitalization

In many ways revitalization is central to our efforts at developing new directions in mathematics at two-year colleges. Without a confident, motivated and committed faculty, efforts at modernizing and refining content will fall well short of what would be desired and of what could be achieved.

In 1981 the Conference Board of Mathematical Sciences reported alarming trends in the teaching environment in two-year colleges.[1] It indicated substantial increases in teaching loads of faculty, noted that larger numbers of faculty were electing to teach overload courses to improve their financial status, and indicated that the number of full-time faculty was decreasing while student enrollments and the number of part-time faculty were increasing. Simply stated, mathematics faculty in two-year colleges was expected to teach more students and exercise more responsibility for the content and instructional quality of the courses taught within the program while not receiving equivalent increases in remuneration. The economic uncertainties and restricted government budgets for education that have existed in the years since that CBMS study do not portend a change in this pattern. It is not surprising that more and more faculty express concern over the even more limited time they have available for expanding their mathematical knowledge and for expanding or modifying their courses. Yet, such efforts by two-year college faculty are essential if they are to retain the instructional quality of their courses.

The economic conditions of recent years have also precluded faculty from relocating to other colleges, primarily because of the substantial expenses of relocating. This means that faculty who in 1979 averaged over 9 years of teaching experience in two-year colleges[2] are today likely to average over 14 years of two-year college teaching experience with much of it at the same two-year college. Many of these instructors will not have had the opportunity of exchanging ideas with new colleagues in new and different teaching settings. Thus, one path of access to revitalization has been restricted substantially.

Not only are the faculty unable to relocate, but they are often working overload schedules and teaching summer courses. Not only does this heavy teaching load limit opportunity for creativity and additional study, but it almost eliminates time for revitalization. Thus, it is not surprising that faculty according to an article in the _Chronicle_ of Higher Education[3] are becoming victims of burnout and the resulting psychological self-protections. These burdens are having a negative impact on the vitality of faculty and on the quality of their instruction.

Traditional methods of protection from faculty burnout, such as sabbatical leave and professional development programs, unfortunately, are some of the first areas for which funding is reduced or eliminated in times of constrained financing. An additional problem is that the criteria for sabbatical leave is narrowly defined. At a number of two-year colleges, including all those in Virginia, sabbatical leave can be used only for obtaining advanced degrees.

The financial consequences for the faculty member who does embark upon a broadening of knowledge by sabbatical leave or leave without pay can be shattering. While those who remain, continuing as usual, receive a salary increase for the year's additional experience in Virginia, the adventurous faculty member who will return refreshed, revitalized and full of new ideas will receive less than half the yearly increase, if any. At a time when faculty need to be revitalized and renewed, two-year colleges, and not just those in Virginia, retain policies which discourage rather than encourage faculty revitalization.

There is an urgent need for new approaches to revitalization of faculty. Our present efforts are too limited and are often inappropriate at a time when faculty in ever increasing numbers need support and uplifting.

It is interesting to note that in China, a less developed country than ours, this commitment to maintaining the vitality and the relevancy of mathematics faculty is strong. In addition to being assigned to only two classes a day, faculty are expected to participate in their own professional development by expanding and renewing their knowledge. In addition, faculty are expected to spend one-half day per week in formal group meetings for the exclusive purpose of the professional development of the group. Such a commitment of time by educators in this country coupled with administrative encouragement and financial support within the two-year colleges would be at least a first step in faculty revitalization.

The Need For Retaining Instructional Quality

In the 1970s and early 1980s, mathematics content has undergone an extensive reevaluation, especially the first two years of undergraduate study. Much of this effort at reevaluation was hastened by inexpensive hand-held electronic calculators and by microcomputers. These have led to expanded emphasis on discrete mathematics, statistics, real data analysis, and "real life" problems, and expanded interest in the use of computers within instruction. Reevaluation has also been fostered by the alarming numbers of students requiring remediation and by the lack of problem solving skills on the part of high school graduates.

The reevaluations of the mathematics curriculum must include occupational-technical courses as well as transfer courses. Revised instructional approaches have been necessitated by the impact that modern technology has had on altering the mathematical needs of technology, business, and health careers. When one looks at how the modern office has changed in the last ten years, it becomes clear that a need exists to approach instruction in these careers in an entirely different way. Also the advances within the electronics industry over the last decade have led one electrical construction company president to indicate that his business is shifting from one which utilizes algebra to one that requires a knowledge of the concepts of Fourier transforms and series.

Even the most basic course in mathematics is not immune from reexamination. This was most evident in the recent conference, "Mathematics K-12: What Is Still Fundamental and What Is Not."[4]

For two-year college mathematics faculty it is essential to be involved in these reevaluations of content and to be knowledgeable about the topics under consideration. Two-year colleges occupy a pivotal location between high schools and four-year colleges and universities. In that position the mathematical content must build upon what has been taught in high school and at the same time serve as a sound foundation for advanced undergraduate study. To have high school and four-year college faculty change their mathematical content while two-year college mathematics content remains static will mean a loss of quality and credibility as well as a loss of student enrollments.

Some faculty will find themselves well-prepared for incorporating the changes into their mathematics classrooms. Some others will find themselves incapable of adapting. Mathematics departments must be prepared to deal with such situations, even to the point of dismissing the non-adaptable faculty members. The largest group will be those faculty who with careful assistance will be able to adapt their instructional approach to incorporate any new directions. It is this last group that will need the focus of retraining efforts if recommended changes in content or approach are to be successful.

Previously, the appropriate knowledge was disseminated by word-of-mouth, through new textbooks, or by presentations at special meetings. In recent years the textbook has not been as good a vehicle for dissemination of new ideas because, for financial security, editors tend to only publish books which reflect the content of courses that currently exist and are well established. It is gratifying to note that there has been a 25% increase from 1975 to 1980 in the number of individuals who have attended and delivered professional presentations.[5]

These efforts at knowledge dissemination have been satisfactory but hardly sufficient. That is supported by the overwhelming popularity of the newly instituted minicourses at national meetings of the Mathematical Association of America (MAA) and the summer workshops sponsored by some of the regional sections of the MAA. These workshops, most of which have had capacity enrollments, have emphasized such topics as problem solving, exploratory data analysis, microcomputers, computer software, mathematics in industry, and numerous computer programming language workshops.

It is crucial that two-year college mathematics faculty be encouraged to participate in such exchanges of relevant knowledge. Workshops and minicourses are an expedient and inexpensive method through which mathematics faculty can learn about areas of interest within the mathematics community.

Another approach to assisting mathematics faculty to retain relevancy is an on-going effort to decrease the critical shortage of computer science instructors by retraining mathematics instructors. The Institute for Retraining in Computer Science at Clarkson College is sponsored jointly by the MAA and the Association for Computing Machinery (ACM). During a two summer program, the institute prepares mathematics instructors to teach computer science courses at the undergraduate level.

These efforts, however, have not been sufficient in light of the multitude of changes being considered in beginning undergraduate mathematics. The concept of the Retraining Institute should be extended into other areas of mathematics. Of course, such institutes should be provided with adequate funding.

The retaining of instructional quality by two-year college faculty also requires an understanding of the necessary mathematical skills and knowledge required within the occupational areas that career oriented students pursue. The opportunity for mathematics faculty to spend a summer or part of an academic year within appropriate industries should be encouraged, supported, and facilitated.

The successes of the MAA minicourses should be shared with those who find it difficult to attend national MAA meetings. Repeating these minicourses in conjunction with MAA section meetings or at meetings of other professional organizations in other parts of the country would provide faculty with an opportunity to participate at locations much closer to their colleges. Some of the minicourses could even take on the characteristics of a touring road show with presentations made in cities across the country.

Retaining quality in instruction, even in the best of times, is an ongoing, time-consuming activity. With the likely advent of a variety of recommended changes and adjustments in the mathematics content of two-year college courses, the efforts by mathematics faculty to retain quality in their instruction will require all the cooperative efforts that colleges and professional organizations can provide. The present efforts have been beneficial but much more remains to be done.

The Need For Quality Instruction For A Changing Clientele

Two-year colleges are facing the prospect of providing instruction to a changing clientele. This will require mathematics faculty to become concerned with providing relevant instruction to this changing student population.

Ten years ago, while four-year colleges and universities were emphasizing higher education for the well-prepared 18 year old on a fulltime basis during the traditional work day, two-year colleges were offering their courses and programs to the underprepared student, the older returning student, the part-time student, and the student needing nontraditional class times and nontraditional modes of instruction. Today, many senior institutions, because of shrinking numbers of traditional students, have also begun to teach these non-traditional students. Although the remediation programs have long been considered the specialty of the two-year college, some of the remediation programs at the senior institutions are considered today to be the most innovative and successful.

During the last decade, two-year colleges also surfaced as the institution which would provide Southeast Asian refugees with an opportunity to succeed in their new country as well as to aid in their assimilation. This role will be of less significance as the flow of such refugees continues to shrink dramatically. However, the two-year colleges will continue to have a role in aiding in the assimilation of increasing numbers of Hispanics from Central and South America.

In the 1970s there was a large pool of adults eager to take enrichment courses offered by two-year colleges. Today, that pool has reduced in size as more and more adults have exhausted the course offerings available at the beginning undergraduate level or have concluded that pursuing higher education is of little interest to them.

Of further impact on the mathematics programs at two-year colleges is the concern in state legislatures and by governors about the extensive cost of remediation in higher education. In Virginia the governor suggested that remediation was not an appropriate activity for higher education.

In California a special task force was established to examine the role of remediation in higher education.

There is a substantial number of high school graduates who have received less than an appropriate high school education in mathematics. With the increasing number of states which are requiring more stringent course work in mathematics from high school graduates[6] two-year college mathematics departments need to be preparing for the future decline in the remediation function as well.

Mathematics faculty must also be concerned about retaining a relevant approach to instruction in light of the changing clientele. Two-year colleges are getting increasing numbers of students from local industries which require retraining on the part of their employees. This retraining is the result of the expansive alterations within the last decade in industrial operating procedures, in the technology that is used, and in the level of mathematical training that is required by employees to perform their altered jobs.

Two-year college mathematics faculty are also going out into industry to assist in retraining of employees. These faculty need to know that they are there in the industry at the discretion of the employer. The employer and employees expect the instruction to be job-related. It is important to note that employees do not have time for content which they fail to view as relevant to their employment.

This approach to industrial training requires a mathematics instructor who can adapt instruction to meet the particular needs of the industry. It also requires a college administration which will provide appropriate acknowledgment and credit for the efforts of the faculty member providing the industrial retraining. Such efforts by faculty and administration will build strong credibility in the community and will provide the college with a continuing supply of students as industries continue to evolve and require additional retraining of employees.

In Summary

The future direction of mathematics instruction in the two-year college will rest on the success of efforts at retaining quality instruction by faculty as content and student needs change and of efforts to retain the vitality of faculty as they continue to become more experienced.

The notion of vitality of faculty appears to be the key link upon which any new curriculum recommendations will depend. There needs to be a new direction on the part of colleges in terms of workload, remuneration, and the role of faculty at the college. Part of this new direction should include a total reexamination of the true cost of providing a quality two-year college education. At the least, colleges need to have enlightened approaches to sabbatical leave and to fund programs in professional development and retraining. These programs, after all, were originally implemented as preventative maintenance to ensure a continued program of quality instruction. Mathematics faculty will soon be confronted by substantial changes in mathematics content at the beginning undergraduate level. Already increased emphasis on calculators, computers, educational software, statistics, problem solving approaches and algorithmic instructional modes are making inroads into altering the variety of approaches to math instruction in two-year colleges. Colleges should well consider immediately investing in preparations for these imminent transitions.

The retaining of quality instruction also becomes vital as mathematics content changes are considered and as student needs and clientele change. Retraining opportunities for faculty need to be intensified, broadened and expanded. Workshops and minicourses need to be offered with more frequency and in convenient locations. Only a properly prepared and revitalized mathematics faculty will be sufficiently equipped to meet the changing needs of students and industry.

Efforts at retaining quality instruction and faculty vitality will require the combined support of all facets of society. The educational community, industry, government and foundations must provide opportunities for and adequate funding for involvement in continued and frequent faculty development. Professional organizations must provide activities and assistance for faculty. College administrations must encourage, support, acknowledge and reward faculty who are involved in efforts at retaining instructional quality and vitality.

In the end, relevant, vibrant mathematics instruction rests squarely on the instructor. Faculty, through efforts at faculty growth and development, must commit themselves to retaining and enhancing the quality of their instruction and their vitality as educators. For only with a vibrant, knowledgeable mathematics faculty will the recommendations of this conference become a reality.

NOTES:

1. James T. Fey and others, *Undergraduate Mathematical Sciences in Universities, Four-Year Colleges and Two-Year Colleges*, 1980-81, Report of the Survey Committee, Conference Board of the Mathematical Sciences, Volume VI (Washington, 1981), pp. 108-111.

2. Robert W. McKelvey and others, *An Inquiry into the Graduate Training Needs of Two-Year College Teachers of Mathematics* (Missoula, MT: Rocky Mountain Mathematics Consortium, 1979), p. 46.

3. *The Chronicle of Higher Education*, March 24, 1982, p. 1.

4. *Mathematics K-12: What is Still Fundamental and What Is Not* (Washington: Conference Board of the Mathematical Sciences, 1983).

5. Fey and Others, p. 110.

6. *Focus* (Washington, Mathematical Association of America), September-October, 1982.

(DISCUSSION BEGINS ON P. 423.)

THE ACADEMIC TRAINING OF TWO-YEAR COLLEGE MATHEMATICS FACULTY

by

Calvin T. Long

SUMMARY

This article discusses the academic training of two-year college mathematics faculty. Courses of study are proposed for baccalaureate and master's degrees for such faculty and it is proposed that the basic requirement for two-year college faculty be the master's degree. The structure of an existing Doctor of Arts degree is also discussed and it is suggested that two-year college faculty, whatever their background, should be involved in continuing education activities on a regular basis.

1. INTRODUCTION

The two-year college system in America is essentially unique. Other countries have educational institutions with goals that overlap those of America's two-year colleges, but they tend to have a more limited focus and are not nearly so ubiquitous as two-year colleges in this country. Indeed, these facts are, to a large extent, the *raison d'etre* of the present conference. The rapid growth of the two-year college system and the wide diversity of goals among the various schools makes it clear that their mission is not well defined. Hopefully, one result of the present conference will be a clarification or redefinition of the appropriate role and mission of American two-year colleges that will be helpful not only to existing institutions but also to those to be established in the near future.

On the one hand, the diversity in the present system is not necessarily bad since a diversity of important educational objectives need to be and are being addressed. On the other hand, this same diversity makes it difficult to know how best to prepare faculty to staff two-year institutions. Preparing teachers to teach the mathematics required in a two-year transfer program is quite different from preparing teachers to deal with masses of students requiring instruction in remedial mathematics or requiring instruction suitable for preparation for various vocations. This poses a difficult problem for institutions preparing future two-year college faculty as well as for the future faculty themselves. What training is the most suitable? How can the needs of these future faculty best be met? I am not at all sure that I have the answer, but I at least will address the problem in what follows. Hopefully, some of the suggestions will have merit; at least they can serve as a point of departure for further discussion.

While this paper deals primarily with preservice preparation, at least brief mention will be made of the in-service or continuing education needs of two-year college faculty. The very title of this conference, "New Directions in Two-Year College Mathematics," is recognition of the fact that the mathematics needs of two-year college students are changing and this clearly implies a need for reorientation and updating of present faculty.

2. UNDERGRADUATE PRESERVICE PREPARATION

As noted in the study of McKelvey, Albers, Liebeskind, and Loftsgaarden [1], in excess of 60% of present two-year college faculty have had extensive experience as high school teachers. In view of the remedial responsibility of two-year colleges and, to a large extent, their role in vocational training, the high school experience does not seem inappropriate. Students in these programs have special needs, and extensive experience in high school surely provides insights and develops skills that allow faculty to better deal with these needs. In any case, the undergraduate training of prospective two-year college teachers should at least meet the requirements for high school teachers detailed in the recent recommendations of the CUPM Panel on Teacher Training of the Mathematical Association of America [3]. In brief the requirements in mathematics would include the content of the following thirteen courses where a "course" is defined as one of three semester-hours duration:

- i) Discrete Mathematics
- ii) Calculus Sequence (three courses)
- iii) Introduction to Computing
- iv) Mathematics Appreciation
- v) Linear Algebra
- vi) Probability and Statistics
- vii) Number theory
- viii) Geometry
- ix) Abstract Algebra
- x) History of Mathematics
- xi) Mathematical Modeling and Applications

Detailed descriptions of these courses can also be found in the _Recommendations_ cited. (See also the guidelines of the National Council of Teachers of Mathematics [2], especially page 11.)

Beyond this, the _Recommendations_ go on to state that "teachers of calculus should have additional work in analysis including the background for teaching the content of the Advanced Placement Calculus." This is to say that teachers of calculus should have had differential equations, and at least one course in advanced calculus.

Finally, with regard to content, I would recommend that the undergraduate program include at least one course in heuristics and problem solving. In many ways, this may be the most important requirement of all. If we can teach students the art of problem solving or, as someone has said, if we can develop thoughtful mathematical behavior in students, then all other mathematical goals will be much more easily attained. After all, the only reason for doing mathematics (of whatever variety) is to solve problems. That is what mathematics is all about!

Of course, the above recommendations only deal with mathematics content which, while necessary, is not sufficient. In addition, two-year college teachers should be broadly educated generally and should also receive instruction in the craft and art of teaching. Thus, in addition to specific knowledge of mathematics and a broad general education, two-year college teachers should be familiar with various approaches to teaching mathematics, with learning theory, with evaluation, with diagnosing and, particularly, remedying student deficiencies in the learning of mathematics. In order for these competencies to be attained it is imperative that institutions provide appropriate course work. In particular, provision must be made for the application of these studies to mathematics instruction through suitable courses in mathematics education including at least one course in methods of teaching mathematics at the two-year college level [see 3, pp. 12-13].

3. GRADUATE PRESERVICE PREPARATION

While what is outlined above is substantial, it is necessarily inadequate. The minimum degree requirement for two-year college mathematics teaching should be a master's degree. The degree could have any one of several designations indicated by the abbreviations M.A., M.S., or M.A.T., all in mathematics, or even M. Ed. But, in any case, like the baccalaureate degree just described, it should include a heavy concentration in mathematics content and also additional course work in professional education. I would recommend an additional course in problem solving, a second semester of linear algebra, a second semester of advanced calculus, a second course in discrete mathematics, advanced differential equations, applied mathematics and/or mathematical modeling, computing, advanced methods of teaching mathematics, a master's special project, and an extensive internship at a two-year college. The master's special project could be pedagogical but it might also be an indepth study of some particular topic like simple continued fractions or recurrence relations. Here students should be expected to dig out material on their own with little assistance from an instructor and also to write up a reasonable account of what has been learned. This last is particularly important. Two-year college faculty need to be confident of the fact that they can read and learn new mathematics on their own, and they also need to develop skill at careful exposition.

Finally, the additional methods course (properly constructed and effectively taught) and particularly the internship (properly supervised) should go a long way toward perfecting the prospective faculty member's skill as a teacher and enhancing his/her understanding of the appropriate role of the two-year college teacher. Beyond that, this experience as much as any other, should help to instill in the candidate a sense of mission and a high degree of professionalism.

Incidentally, those students who decide to go into two-year college teaching after dropping out of Ph.D. programs should be required by hiring institutions to submit credentials equivalent to what has just been detailed for the undergraduate and master's training of such faculty. In particular they should be required to take undergraduate and graduate courses in learning theory, evaluation, diagnosing and remedying student deficiencies, and methods which they would not normally take otherwise. They should also be expected to complete an internship in a two-year college as discussed above.

4. THE DOCTOR OF ARTS DEGREE

As reported by McKelvey, *et al.* [1], an increasing number of two-year college teachers are seeking the doctorate. This is all to the good. But the degree should be especially designed and should probably be something other than the Ph.D. A number of graduate schools have settled on the Doctor of Arts designation while others use Doctor of Mathematics Education, Doctor of Science Education, or some form of the Ed. D. Whatever the designation, the degree should have a heavy concentration of mathematics and mathematics education courses as opposed to courses in professional education.

The aim of the degree program should be to produce a highly professional teacher of two-year college mathematics who is broadly knowledgeable in mathematics and is also an expert practitioner of the craft of teaching in all its aspects as it applies to two-year colleges. Such a teacher should be a leader on his/her campus, should be involved in professional organizations, and should be continually involved in devising new and more appropriate curricula and teaching strategies. Such a teacher should also be actively promoting the profession and publishing as appropriate in suitable journals.

Programs leading to the Doctor of Arts or similar degrees exist at a number of institutions. On my own campus, it is a high level degree similar to the Ph.D. and yet with notable exceptions. Briefly, the requirements are as follows:

i) Forty-five hours of graded courses including nine hours of combined linear algebra and analysis (advanced linear algebra, calculus on E^n, basic complex analysis, harmonic functions, metric spaces, integration, Hilbert spaces, Fourier and Laplace transforms, special functions, and group representations) and twenty-four hours chosen from among a large selection of graduate level courses in pure and applied mathematics and statistics.

ii) One semester courses in numerical analysis and foundations of mathematics, a two-semester seminar on the history of mathematics, and a seminar in undergraduate mathematics instruction.

iii) Two advanced courses in a department other than mathematics are recommended; the courses should draw heavily on mathematics and illuminate its place in science and society.

iv) A course in computer organization and programming, one in program design and development, and at least one more advanced computing course. Computing is expected to be a continuing and vital part of the mathematical repertoire of every candidate for the Doctor of Arts degree. Particular emphasis should be placed on the use of small computers and other aspects of computer science most likely to be useful to a teacher at a two-year college.

v) Advanced educational psychology.

vi) An advanced education course concerned with the history, sociology, and philosophy of higher education. Possibilities on our campus include The Community and Junior College, Undergraduate and Community College Teaching, and The American College and University.

vii) A minimal proficiency is required in two foreign languages--French, German or Russian.

viii) Each DA student is required at some time during the program to be responsible, under supervision, for teaching at least one undergraduate mathematics class for a semester.

ix) The student is strongly recommended to spend at least one quarter as an intern (i.e., as a full-time teacher) at a two-year college under close supervision by someone in that college and by a member of the faculty at Washington State University.

x) A thesis is required. This may embody an original contribution to mathematics, but normally will be a scholarly dissertation which could be historical, critical, or philosophical

> in nature, or deal with suitable problems in mathematics education. In the latter case, the emphasis will be on mathematical scholarship. Studies devoted primarily to surveys, curriculum construction, experimental teaching techniques, and the like will probably not be acceptable. The thesis should exhibit considerable learning, insight, and skill in exposition.

Our department has never pushed the Doctor of Arts program with vigor and only one student, currently teaching in a four year college, has completed the degree thus far. However, the department is now giving the degree increased emphasis and six students, some of whom are contemplating careers in two-year colleges, are presently enrolled in the program. My own opinion is that our degree is perhaps too near the Ph.D. and that it might well include less ultra-sophisticated mathematics which has little relationship to what goes on in two-year colleges. To put it the other way around, it might well include more course work in curriculum development, on teaching strategies--particularly for remediation and for teaching vocational mathematics, on problem solving and the teaching of problem solving, and on applications relevant to the two-year college scene. Also, significant theses devoted to curriculum development and to the development of new teaching strategies ought not be viewed as exceptional as they presently are.

5. CONTINUING EDUCATION FOR FACULTY

There are many reasons why continuing education or continuing development activities are a must for two-year college faculty (as, indeed, they are for teachers of every level--K through university).

In the first place, increasingly rapid change, orderly and disorderly, haphazard and reasoned, is the hallmark of our times and the implications of this fact are obvious to us all. In order to keep abrease of the times, teachers at every level need to be heavily engaged in seminars, workshops, short courses, conferences, and professional meetings. We also need to be reading a seemingly endless array of professional papers, pamphlets, books, and periodicals. How else can we be appraised of the latest developments in mathematics, in the teaching of mathematics, in the application of mathematics, in computing and the relationship between computing and the subject matter of mathematics, and in the use of computers in the teaching of mathematics, etc., etc.? The need to know and to keep abreast is dramatic and compelling. There is no denying these facts. To attempt to hold the line is, in fact, to fall hopelessly behind.

But there is another, and it seems to me, even more compelling reason for teachers to study and to continue to expose themselves to new ideas in their particular area of interest and expertise. Teaching is at once a satisfying and enriching task and also an extremely enervating one. Good teaching requires the regular expenditure of enormous amounts of nervous, mental, and physical energy, and we are all aware of the large numbers of teachers who leave the profession because of "burn out." Differently put, good teaching requires the maintenance of a high level of enthusiasm on the part of the teacher. Like the old cliché of leading a horse to water but being unable to make him drink, it has been said that it is impossible to teach a student anything. On the other hand, it *is* possible for an adroit and enthusiastic teacher to inspire a student to want to learn and to do so successfully. This is at once the task, the challenge, and the reward of the teacher.

In short, I claim that we teachers can only maintain our zest for mathematics and our ability to inspire in our students a desire to learn mathematics, by continually investigating new topics in mathematics *for ourselves*. Not because we expect to turn around and present these ideas to our students, not because such study is necessary if we are to keep abreast of the times, but because such activity is essential for the maintenance of our own personal mathematical well-being, because we want to allow ourselves continually to be confronted afresh with the unexpected delights and subtleties of new mathematical ideas. If you please, what I am saying is that we teachers of mathematics continually need mathematical enrichment experience *for ourselves* and that obtaining such experience is worth the expenditure of a considerable amount of time and effort. There, of course, is the rub; for we are all only too well aware of the demands modern life makes on our time and energy. But the prize is worth the cost and, if we are to stay alive and active as teachers, the effort simply must be made.

The Indian philosopher Tagore summed the matter up very well when he penned the following words:

> A teacher can never truly teach unless he is still learning himself. A lamp can never light another lamp unless it continues to burn its own flame. The teacher who has come to the end of his subject, who has no living traffic with his knowledge but merely repeats his lessons to his students, can only load their minds; he cannot quicken them. Truth not only must inform, but must inspire. If the inspiration dies out and the information only accumulates, then truth loses its infinity (and the teacher loses his effectiveness).*

REFERENCES

1. Robert McKelvey, Donald Albers, Shlomo Libeskind, Don Loftsgaarden, An Inquiry into the Graduate Training Needs of Two-year College Teachers of Mathematics, The Rocky Mountain Consortium, 1979.

2. Guidelines for the Preparation of Teachers of Mathematics, National Council of Teachers of Mathematics, Reston, 1981.

3. Recommendations on the Mathematical Preparation of Teachers, The Mathematical Association of America, Washington, 1983.

*Parenthetical expression added by the present author who has also lost the reference on this quotation. If any reader knows it, please notify the author.

(DISCUSSION BEGINS ON P. 423.)

CURRENT CONTINUING EDUCATION NEEDS OF TWO-YEAR COLLEGE FACULTY MUST BE MET !

by

Karen Tobey Sharp

SUMMARY

CURRENT CONTINUING EDUCATION NEEDS OF TWO-YEAR COLLEGE MATHEMATICS FACULTY MUST BE MET!

Rapid developments in technology and changing student enrollment patterns make updating of two-year college mathematics faculty (TYCMF) skills imperative.

Furthermore, retraining of the current staff is feasible. Since approximately one-half of the faculty is under 45 and can expect to teach for at least another twenty years, efforts expended to update current faculty can have long-term benefits to the institutions involved. A substantial portion of TYCMF attempt to update their skills themselves by enrollments in graduate courses and attendance at conferences.

In the past, faculty have preferred short, intensive courses, sabbatical year formats or summer session programs. Current formal advanced degree programs do not fill the need for TYCMF renewal.

Continuing education needs of TYCMF should be met through expansion of government funding, endeavors by professional mathematical societies, industry and business support, and college and faculty action.

The consequences of lack of action in TYCMF continuing education are grim. There will be further inroads into two-year college programs, such as has occurred in remedial mathematics since four-year colleges started offering such courses. It is more efficient and less expensive to retrain and update TYCMF than to create or utilize other institutions whose design might, in the long run, be less effective than the two-year college has been.

CURRENT CONTINUING EDUCATION NEEDS OF TWO-YEAR COLLEGE MATHEMATICS FACULTY MUST BE MET!

Professor Karen Tobey Sharp

I. RAPID DEVELOPMENTS IN TECHNOLOGY MAKE UPDATING OF TWO-YEAR COLLEGE MATHEMATICS FACULTY SKILLS IMPERATIVE.

A crucial need for continuing education of two-year college mathematics faculty (TYCMF) is not hard to show. In fact, instituting methods of renewal for K-12 math teachers was a key recommendation of the report, Educating Americans for the 21st Century, by a National Science Board Commission of the National Science Foundation (NSF). The need for updating TYCMF is just as great and can be found in at least four areas---content, curriculum, methodology and attitude.

As a result primarily of the rapid developments in technology, notably computers, both content taught at the two-year college in mathematics and approaches to teaching mathematics should undergo considerable change. Many TYCMF do not have skills to handle such approaches and courses.

As indicated in the 1983 report, "New Goals for Mathematical Sciences Education", by the Conference Board of Mathematical Sciences (CBMS), "The challenges and opportunities arising from computer science will have a significant impact on mathematical sciences education....The content of traditional courses such as linear algebra and calculus will be affected by computing. Computers will make several new mathematical science courses such as discrete mathematics and mathematical modeling of great importance, and new faculty will routinely use computing in their research and teaching. Thus, it is critical for efforts in renewal of collegiate mathematical sciences faculty to develop awareness of computers and the mathematical methods they imply." The report further states, "Collegiate faculty will also need programs to increase their knowledge of...problem-solving of the ill-posed, real-world sort,...the mathematics of technology studies, and t many new examples of applications relevant to service courses in areas jus beginning to use mathematics as an important tool."

In a study funded by NSF and the Rocky Mountain Mathematics Consortiu "An Inquiry Into the Graduate Training Needs of Two-Year College Teachers of Mathematics," the statement is made that the current situation in two-year colleges points to the "...clear need for a regular and sustained program of continuing education for two-year college mathematics faculty."

The biggest deterrent to providing mathematically competent citizens our country is the lack of qualified mathematics teachers. Perhaps even the common defense and general welfare of our country is threatened by thi situation (Willoughby, 1983).

A 1981 CBMS survey of undergraduate mathematics indicates that in two-year occupational/technical programs, the number of students now excee college transfer enrollments. Since 1975, computer course enrollments hav exploded and now outnumber those in calculus. "Building on a small base i 1975, computing courses jumped by 850%!" (McKelvey, 1979). Access to computers is up sharply, but the use of computers in mathematics teaching has increased little since 1975. The growth in remedial course enrollment has slowed, but still amounts to 42% of two-year college mathematical science enrollments.

Further, William C. Missimer, Jr., Executive Vice-President, Pratt & Whitney Group, United Technologies Corporation, states that, "Within three years, there will be a 31 percent increase in demand for technical graduates." He further states, "Colleges are finding it difficult to meet industry's demands for engineers because many of the schools don't have sufficient faculty....In computer science, where there is a burgeoning need, 16 percent of the faculty positions are going begging."

As a result of this new content, growing out of the rapid development of modern technology and changing patterns of student enrollment, there is a vast need for curricular change. As stated in "New Goals for Mathematical Sciences Education," the CBMS report, "The fundamentals of mathematics desirable for students at...college levels have, in the view of many mathematics educators, changed radically, yet the changes are not reflected in core curricula." As further evidence of the need for curricular reform, the National Science Board report, Educating Americans For the 21st Century, recommends "...improving the quality and usefulness of the curriculum." When changing the curriculum, the Commission recommended that there be a focus on all students, not just pre-professionals in science and mathematics. While this recommendation applies specifically to K-12 mathematics, the two-year college currently teaches all of grades 9-12 mathematics. Changes such as are recommended by this commission must therefore be reflected in the two-year college curriculum as well. Faculty need to be informed regarding what curricular changes are needed and how to implement them.

Better methodology in teaching is a third way in which TYCMF need updating. In a recent survey, teachers indicated that their biggest problems were teaching unmotivated students, having to cover much more material than students can absorb, and coping with the deluge of remedial mathematics (McKelvey, 1979).

Finally, any comprehensive continuing education program must incorporate a consideration of the teaching climate and the attitudes of the TYCMF.

Performance of teachers is a function of ability, motivation and climate. Little has been done in staff development beyond ability (Hammonds, 1982).

Traditionally the administration has found it easier to employ new instructors to perform new or different functions than to retrain old instructors. This approach worked as long as the expansion of the two-year college was rapid. Now, with the slackening of growth, there is a need for staff development (Cohen, 1982).

II. RETRAINING OF CURRENT STAFF IS FEASIBLE

Such retraining and development of TYCMF is feasible. Since approximately one-half of the faculty is under 45 and can expect to teach for at least another 20 years, efforts expended to retrain and update current faculty can have long-term benefits to the institutions involved. As employment of part-time faculty continues to sky-rocket, hiring of newly trained, young faculty in full-time positions seems unlikely.

In 1980, 59% of the faculty reported attendance at one or more mathematics conferences per year, while 22% said they were taking additional graduate courses that year (Fey, 1981). So the desire and the ability to learn is certainly there (McKelvey, 1979).

III. COMMITMENT SHOULD BE THE UNDERLYING THEME FOR ALL CONTINUING EDUCATION ENDEAVORS.

If the needs for updating TYCMF are to be met, a variety of approaches must be diligently pursued, and the underlying theme of all such endeavors must be commitment and cooperation on both the part of the planners and the faculty. In "New Goals for Mathematical Sciences

Education," CBMS speaks of "...the need for a continuing commitment from all parties participating in renewal programs...."

The administration of the colleges must exhibit an interest in what is most important to the faculty---the learning of their students. Research indicates that almost every factor influencing individual development, productivity, and creativity is fostered by management (Duncan, 1982).

Stephen S. Willoughby, President, National Council of Teachers of Mathematics (NCTM), stated in January of 1983 that if a true national commitment were to be made to education, then several important actions should be taken. Notably he recommended providing "...more opportunity and incentive for teachers to continue their professional development through participation in activities of professional societies and through further formal education." He further recommended standards for becoming and remaining a teacher be improved and that the salary of every teacher in the country be doubled. While this last recommendation seems drastic, the recent Carnegie Foundation report noted that teachers are now among the nation's lowest paid professionals, averaging $20,500 per year in salary. Many jobs in industry that do not require a college education pay more than that (Missimer, 1984).

IV. TRADITIONAL CONTINUING EDUCATION FORMATS.

Lessons learned in the past can be a guide in formulating plans for the future. Traditionally, faculty development has taken the form of discipline-based institutes, release time, sabbatical leaves and tuition reimbursements for instructors to attend university-based programs. There have also been short courses or workshops on pedagogy sponsored by single institutions or by institutional consortia. The instructors have preferred courses and programs in their teaching field, offered by universities close at hand, that enabled them to gain further knowledge in their field and to rise on the salary scale (Cohen, 1982).

The most popular formats have been short, intensive courses of three to five days duration, sabbatical year formats, and summer session programs (McKelvey, 1979). An apparent concern, reflected in all these preferences, is the need to protect the faculty member's income while at the same time updating skills. This need must be considered in plans to update skills so that the faculty is freed from the worry of financial loss during retraining.

While participating in a formal degree program can provide a focus and a systematic approach to continuing education, obtaining a standard doctorate is termed "unimportant" to a majority of TYCMF (McKelvey, 1979).

Current advanced degrees available (D.A., Ph.D., Ed.D) are not likely to meet the needs of TYCMF even though obtaining such a degree is one way to advance both professionally and monetarily. The most obvious reason for such an attitude on the part of a majority of the faculty is that such programs have neither the content nor the methodology needed in two-year college mathematics instruction today. If formal degree programs are to have a place in the retraining of TYCMF, then vast reorganization of the content of those degrees must take place.

V. RECOMMENDATIONS FOR CONTINUING EDUCATION PROGRAMS.

A. EXPANSION OF GOVERNMENTAL FUNDING

Educating Americans For the 21st Century proposes solutions for the improvement of K-12 mathematics education. The report states, "Most of the nation's 200,000 secondary mathematics and science teachers...

require additional training because of the rapid development of new knowledge in mathematics and science...." To help with the educational needs of current K-12 mathematics teachers, the Commission recommended that NSF establish state-wide or regional teacher training programs using the new information technologies such as telecourses and interactive telecommunications. In addition, it was recommended that the states develop teacher training and retraining programs with colleges, universities and museums. Further, the states should develop regional training and resource centers which would include computer instruction and software evaluation. These centers could serve as the focus for participation of business and government in education.

Other notable advocates of increased governmental funding of faculty retraining include CBMS, Stephen Willoughby, President of NCTM, and Amber Steinmetz, President of The American Mathematical Association of Two-Year Colleges (AMATYC).

Nearly all of the current funding from NSF, historically the prime source of governmental support for mathematics education, is either in the area of K-12 or in mathematical research. These efforts are strongly supported. However, it is imperative to the health of the two-year college mathematics program that NSF recognize the importance of the two-year college in the educational network of America today. New programs should be instituted to retrain the current TYCMF or access to existing and future programs for K-12 teachers should be encouraged strongly. The National Center for Education Statistics reports that the proportion of students in post-secondary education attending two-year colleges rose from 26% in 1970 to almost 40% in 1982! Yet there are few, if any, programs directed toward the benefit of the two-year college and none in faculty retraining or renewal!

There are strategies that have proven effective in the past that shou be revived and supported. The Chautauqua Short Courses and various stipen for summer institutes and academic year study are prime examples (CBMS, 1983). The NSF institutes of the last two decades were successful program which retrained many teachers in the current skills of the day. More than half of today's TYCMF previously taught in secondary schools and received additional training in those NSF institutes.

B. ENDEAVORS BY PROFESSIONAL MATHEMATICS SOCIETIES.

Faculty retraining is currently being undertaken by the various professional mathematics societies, such as AMATYC, The Mathematical Association of America (MAA), and NCTM.

MAA presents minicourses at its conventions. These minicourses are in many of the areas that have been discussed---such as discrete mathematics and computer technology. AMATYC has been presenting workshops for several years which are held the day before its regular national convention. Topics in which faculty currently need instruction have been prominent, such as computers and applications of mathematics. AMATYC is planning a summer workshop in 1985 to provide TYCMF continuing education carrying graduate credit in discrete mathematics and computer technology. Faculty appear eager to learn. MAA had waiting lists for all its minicourses from 30 to 60 persons at its January, 1984, convention. Many interested faculty were turned away due to lack of space and the overwhelming demand.

CBMS recommends that "...the professional societies in the mathematical sciences, especially NCTM, MAA, and AMATYC, seek support for projects to demonstrate effective models of the various faculty renewal activities...." Further, Educating Americans For the 21st Century recommends that "the professional societies in mathematics should play an active role in curriculum development, review and revision."

These activities by professional societies are commendable but are only a beginning. Continuing education by professional societies is "high value at low cost." (Moneysmith, 1984). However, funding is needed to expand the scope and availability of such endeavors. This additional support should be given by the federal government, perhaps through NSF.

C. INDUSTRY AND BUSINESS SUPPORT.

An exciting and promising source of support for TYCMF development is business and industry. The call for active and aggressive participation from this source is rapidly increasing. Such participation is appropriate and necessary as business and industry are prime recipients of the benefits of quality education. Since faculty skills are rusty and have not kept pace with technological developments, it is necessary in the face of the tremendous need and lack of sufficient resources elsewhere that business and industry assume a major role in the retraining of TYCMF.

As William C. Missimer, Jr. has stated, "Business and industry...must become a focal point to bring about needed improvements in the scientific and technological literacy of our youth....Everything that might go into an attack on the problem must include a high-priority goal: the restoration of competent, fully-certified math and science staffs throughout all levels of our educational system."

He specifically recommends that business and industry should:

> *Help teachers relate classwork to the workplace by hiring them during the school vacations.
>
> *Invite teachers to attend in-plant training programs so they can see how math skills are being applied.
>
> *Send more of their employees to visit schools and discuss the need for mathematics.
>
> *Fund research centers on campuses where faculty, industry and students could study new technologies.
>
> *Provide quality programs of financial support and academic encouragement to help ease the loss of mathematics faculty.
>
> *Donate surplus high-technology equipment to schools.
>
> *Work with teachers to revise curricula so that there is a proper emphasis and balance in the changes.

The idea of industry hiring faculty during vacations has been advocated by others. Stephen Willoughby has said industry should be encouraged "...to hire mathematics and science teachers during the summer to enhance their incomes and their knowledge of how their subjects are used in industry."

The concept of industry and business participation in faculty continuing education is carried a bit further in a model proposed by Richard Afred and Nancy Nash in a recent Community College Review article. They recognize that there is a difficulty in encouraging faculty to update teaching skills if course content or methodology must change. To remedy this difficulty, they proposed that faculty be placed in business and industry settings for six to twelve months to "...learn new skills, validate theory, study current practices, and learn problem-solving techniques." Simultaneously, industry professionals and technicians could be placed in the two-year colleges to teach the load of the staff placed in the industrial or business setting and to evaluate and modify occupational curricula to comply with the changing technology. The benefits of this plan would be the modernization of the college curricula in accordance with emerging technology while simultaneously preparing the faculty member to return to an academic setting to teach the new curricula.

Participation of business and industry in faculty development is important also because TYCMF perceive their background as being least adequate in the areas of application (McKelvey, 1979). Furthermore, the plan is made more feasible by the fact that most two-year colleges already have contacts with business and industry through liaison personnel for the occupational, apprenticeship and technical programs. Added stimulus for this concept could be given by local, state, and federal governments in the form of tax credits and like incentives.

With business and industry contributing to the retraining of TYCMF, all parties involved win. The faculty member becomes acquainted with the latest technology, is revitalized and gains professional enrichments. Students benefit from an improved curriculum and a more knowledgeable teacher. The college benefits from an updated program. Industry benefits from a better trained pool of employees (Conrad, 1982). Given the substantial benefits that business and industry would reap, it is appropriate that they contribute financially to TYCMF continuing education.

D. COLLEGE AND FACULTY ACTION.

Finally, the two-year college itself must recognize the very real and urgent need for development of its mathematics faculty, as well as the fact that the quality of its faculty is ultimately its own responsibility. An active, coordinated effort should be instituted at once by the two-year colleges across the nation to implement the various plans proposed for faculty retraining. The help and support of the faculty, professional mathematical societies, business and industry must be solicited actively. Local, state, and federal grants must be encouraged to recognize the place of two-year colleges in the mathematics and science education of our student population. The various programs already available and which are currently being instituted to retrain K-12 mathematics teachers must be expanded to include TYCMF.

If efforts to expand the base of support are unsuccessful, then the two-year colleges must be prepared to extend their own resources and programs for continuing education for TYCMF. More monies should be provided and more sabbaticals should be awarded in the area of mathematics and computer science to retrain faculty. If the two-year colleges want to remain intact as the type of training institutions they are, then they must make a firm commitment to retraining their mathematics faculty in the modern methods and content.

The individual in many professions is responsible for keeping his or her own skills current. In these professions the consequences of inadequate skills may be lack of advancement or, in extreme cases, loss of employment. Many TYCMF have grown lazy as a result of employment protection provisions of contracts and tenure laws. Such job security is certainly good to have, but this apathy is contributing to many TYCMF rapidly having obsolete skills. To remain a professional, it is necessary for TYCMF to actively pursue updating of their skills as well as being the recipients of support and actions on the part of others.

Particularly in the area of computer training, some TYCMF have resisted learning new skills. The reasons for such avoidance include fearing a requirement to teach computer science rather than mathematics, lacking either the time or financial resources to train in these areas, or resisting the infusion of computer science into the mathematics curriculum. The time is past when mathematicians can be ignorant about computers. Just as resistance to classroom use of hand-held calculators a few years ago is largely a thing of the past, now computers should be regarded as affecting almost all aspects of our lives. For mathematics teachers to remain ignorant regarding computers is to become incompetent.

VI. CONSEQUENCES OF LACK OF ACTION.

The results of continued apathy or lack of diligence in finding a solution to this problem are grim indeed. Declining college enrollments and changing student bodies at four-year colleges and universities have resulted in expansions of programs into areas regarded until recently as the domain of the two-year college. For example, until about three years ago, post-secondary remedial mathematics was taught almost entirely by two-year colleges. Increasingly today, four-year colleges and universities are adding remedial mathematics courses to their curriculum.

Unless the two-year college actively assumes a greater role in seeing that its technical and scientific staff are properly retrained, more erosion of current programs will occur. In addition, the two-year college will be unable to offer a modern course of study needed as a result of our changing technology. If the demands of business and industry for modern programs cannot be met appropriately by the two-year college, then other institutions will be called upon to meet the need. Perhaps the four-year colleges will step into that place, or industry and business themselves will establish their own training programs. The two-year colleges have been in place and functioning across the nation for half a century. Presumedly, it is more efficient and less expensive to retrain and update TYCMF than to create or utilize other institutions whose design might, in the long run, be less effective than the two-year college has been and can be.

One thing is certain. Without modern teaching techniques and competent faculty in mathematics and science, the two-year college will be left behind in a high-tech society.

REFERENCES

Afred, Richard L. and Nash, Nancy S., "Faculty Retraining: A Strategic Response to Changing Resources and Technology," COMMUNITY COLLEGE REVIEW, Fall, 1983, pp. 3-8.

Cohen, Arthur M. and Brawer, Florence B., THE AMERICAN COMMUNITY COLLEGE. San Francisco, CA., Jossey-Bass Publishing Co., 1982.

Cohen, Arthur M., "Work Satisfaction Among Junior College Faculty Members," Paper presented to Annual Meeting, California Educational Research Association, November 28-29, 1973.

Conference Board of the Mathematical Sciences (CBMS), "New Goals For Mathematical Sciences Education," Report of a conference sponsored by CBMS, November, 1983.

Conrad, Clifton F. and Hammond, Martine, "Cooperative Approaches to Faculty Development," COMMUNITY COLLEGE REVIEW, Fall, 1982, pp. 48-51.

Committee on the Undergraduate Program in Mathematics (CUPM), RECOMMENDATIONS ON THE MATHEMATICAL PREPARATION OF TEACHERS. Mathematical Association of America (MAA), 1983.

Duncan, Mary Ellen and McCombs, Carol, "Adult Life Phases: Blueprint for Staff Development Planning," COMMUNITY COLLEGE REVIEW, Fall, 1982, pp. 26-35.

Fey, James T., Albers, Donald J. and Fleming, Wendell H., UNDERGRADUATE MATHEMATICAL SCIENCES IN UNIVERSITIES, FOUR-YEAR COLLEGES, AND TWO-YEAR COLLEGES, 1980-81. Washington, D.C., Conference Board of the Mathematical Sciences (CBMS), 1981.

Hammons, James O., "Staff Development Isn't Enough," COMMUNITY COLLEGE REVIEW, Winter, 1982-83, pp. 3-7.

Hansen, Desna W. and Rhodes, Dent M., "Staff Development Through Degrees: Alternatives to the Ph.D.," COMMUNITY COLLEGE REVIEW, Fall, 1982, pp. 52-58.

McKelvey, Robert; Albers, Donald J.,; Libeskind, Shlomo and Loftsgaarden, Don O., AN INQUIRY INTO THE GRADUATE TRAINING NEEDS OF TWO-YEAR COLLEGE TEACHERS OF MATHEMATICS. Missoula, Montana. Rocky Mountain Mathematics Consortium, 1979.

Missimer, William C. Jr., "Business and Industry's Role in Improving the Scientific and Technological Literacy of America's Youth," T.H.E. JOURNAL, February, 1984, pp. 89-93.

Moneysmith, Marie, "Continuing Education. High Value at Low Cost," A&SM, Apr/May, 1984, pp. 41-44.

National Science Board Commission on Precollege Education in Mathematics, Science and Technology, EDUCATING AMERICANS FOR THE 21ST CENTURY. Washington, D.C., National Science Foundation (NSF), 1983.

Steinmetz, Amber, Santa Rosa Junior College, Santa Rosa, CA., personal interview, January, 1984.

Willoughby, Stephen S., "The Crises in Mathematics Education," MATHEMATICS IN MICHIGAN, Summer, 1983, pp. 4-13.

(DISCUSSION BEGINS ON P. 423.)

DISCUSSION

FACULTY RENEWAL AND PROFESSIONALISM

1. Part–Timers and Professionalism
2. Roots of a New Professionalism
3. Differentiating Among Faculty
4. Professionalism Ex Cathedra
5. Professionalism Versus Money
6. Faculty Exchanges
7. Rank and Tenure?

FACULTY RENEWAL AND PROFESSIONALISM

Tucker: It seems to me that the cadre of full-time faculty might have the additional responsibility of overseeing the part-time sections and not just the scientific sections. This might be a sort of trade-off of less teaching time for more professionalism. The costs of replacing the full-time member with a part-time adjunct is not what it is to hire another full-time person. One may feel that one is selling one's brothers and sisters up the river, or something like this, when one says that there are two tiers and that the part-timers and the full-timers should be treated differently. It seems to me it's a basis for making an argument to administrations that you really have to treat the full-timers as professionals in a special way and that they in return oversee the whole program. They can take an interest in what the part-timers are doing, visit the classes, and so on.

Davis: It's a very good idea, but unfortunately it has already been thought of in terms of the administration's viewpoint. At most institutions that I know, the use of part-timers is already fairly high and climbing higher. It's been a matter of saving money, but there's not been any trade off given to faculty because of that.

Part-Timers and Professionalism

Smith: Nor would I want to imply that we would like any trade off. I would not propose any trade-off even if that were possible. I think the quality of teaching by part-time faculty is lower than that of full-time faculty.

Long: This calls for a recommendation I think from this group to really minimize the use of part-time people. It's a cop-out, as I see it, and it's antiprofessional. It's a cop-out on the part of the central administration of the community college.

Albers: There is some rather hard data available from the 1980 Survey of the Conference Board of the Mathematical Sciences which shows the following. Between 1975 and 1980 enrollments in mathematics in two-year colleges increased by 26 percent. During that same period the number of full-time faculty actually decreased. What was going on? Well, of course, the part-time component increased; but you have to look more closely at that. We discovered that a great many full-time faculty were, in fact, part of the part-time pool. The fraction of faculty teaching overloads was substantial. Reason?--most likely--money.

Case: At their own schools?

Albers: Both at their own schools and at other schools.

Curnutt: I want to say just one more thing in response to Alan's comment. Part-time instructors are not just adjunct faculty, who come in from an industrial setting and teach a special course that has some particular bent. They teach core courses and many of them are just as highly-trained and nearly as experienced teachers as full-time faculty. The only significant difference is that they don't participate as fully in things like campus governance. The really big issue is that they don't get paid well.

Rodi: I found it rather interesting that Alan drew that comparison. I was going to draw exactly the opposite sort of comparison. Why, at universities and four-year colleges, is so much teaching done by graduate assistants? Why doesn't it work the other way? Why don't we go for more full-time people at universities and colleges?

Roots of New Professionalism

Kaput: I think there's a root difficulty having to do with the nature of professionalism and on what professionalism is based. At four-year schools the professionalism that exists tends to be based on the knowledge-production enterprise--research in your particular discipline. That, coupled with the amenity of relatively low teaching loads yields an automatic professionalism. There is a kind of control of our behavior that follows from these things. There is relatively little professionalism at four-year colleges based on the business of teaching. Riesmann has written a beautiful book, analyzing the faculty culture, showing that teaching plays a very small role as a focus of our professionalism. Since there's no research in any particular discipline that's significantly supportive of two-year schools, you're left with a relatively weak basis [instruction] on which to form the professionalism. Art Cohen suggested that somehow a sense of profession be built on the instruction enterprise. I think that's a nice idea, but, unfortunately, we don't have a coherent philosophical and theoretical organization lying behind our instructional decisions. There's no real science of instruction at this point on which you can build a technology and center a sense of professionalism. Over the next decades that will change considerably as we begin to develop a real theory of the science of instruction.

Long: I think that there are many aspects of teaching that can be utilized now as highly professional activities that are very appropriate for junior college people. They should be writing, I think, for suitable journals about special projects and programs at their colleges. They should be devising curricula. They should attend professional meetings and could speak at professional meetings.

Kaput: I think that's true and should be encouraged. The problem is that the kinds of talks that are described are of the type, "My math lab worked well because we did things in this way"; they are not particularly deep and the results described generally do not involve significant differences from one group to another. Once we get a better basis for writing those papers, then in fact, we'll have a better basis for building that type of instructional professionalism.

Warren: We appear to be looking for something external to motivate professionalism. I guess I'm concerned that pride in the responsibility and the job that one has are not incentives enough to be a professional. I said yesterday that I think you people, those of you who are from two-year colleges and are here, certainly represent the top end of the scale of professional, committed people. But for every one of you who are here, I know that you can name someone who you feel is unprofessional by your standards. But, yet, if you come from the same institution, you get the same salary, probably you get the same working conditions, you get the same teaching load, and you've got the same student body. What makes the difference? What is the difference between those "professionals" and those "unprofessionals"? I think Cal Long said it quite well in his paper: "Good teaching requires the maintenance of high levels of enthusiasm on the part of the teachers. Like the old cliche of leading the horse to water and being unable to make him drink, it has been said that it is impossible to teach a student anything." On the other hand, it is possible for an adroit and enthusiastic teacher to inspire a student to want to learn. This is at once the task, the challenge, and the reward of the teacher. I don't think you can take those people who are now unprofessional and give them sabbaticals and give them increased salaries, or anything else and make them professionals. They're enjoying their summers, they're enjoying their second jobs and they want to have the money so that they can continue to do that. It's the new teachers who are coming in who are really expert and committed to teaching. They are the people who are going to make the difference.

Differentiating Among Faculty

Long: I couldn't agree more, Bill. I think that that's where the nub of the problem is, and I guess that what I'm seeing is that the administrations of the two-year colleges need somehow to do a much better job of recognizing those people who are, in fact, professional and those people who are not, and differentiate. And this goes right square in the face of bargaining organizations. But that's where a decision--a determination--has to be made. This person is professional, and that person is not. This person is effective and that person is not. A differentiation has to be made.

Davis: Although I believe that the key to improvement of our situation is faculty going out and building themselves up, I think also we need to have reinforcement from our colleges, from our communities, and from industry. There has been little reinforcement, if any at all, of what faculty do. I have seen good faculty who have come in devoted, dedicated, and, after fifteen years, are beaten down and deadened. And unless we can turn that aspect around, even those new, invigorated people we bring in will eventually become the same kinds of moles that we see.

Professionalism Ex Cathedra

Page: I do not believe that you can mandate professionalism ex cathedra. I think it's very nice for people to pontificate on what educators should do: we should go to meetings; we should attend colloquia; we should do all that. I don't think it's going to work because if an individual's self image is not one of being a professional, nothing you suggest to that person is going to happen. If a person feels like a professional, then that individual is going to seek out means to enhance and reinforce that professionalism. Now, I also agree with Ron that there is a burn-out and a sense of dissolution. Studies have shown that one's threshold for coping is increased depending on how valued one's output is. In particular, if a boss does not appreciate one's work, it is very difficult for that individual to sustain enthusiasm. Two-year college faculty are the unsung heroes.

Next is the matter of locus of control. Now when we went into teaching, we were enthusiastic; we had visions; we had beliefs and ideals that we could make contributions. But then we met the realities and the pragmatics of the situation, the cost economics, and the problem of allocation of resources. There is not enough money to experiment with teaching methodology; there's not enough money to take sabbaticals and learn; there's not enough money to try different processes. In short, there's not enough given to further that kernel of interest and enthusiasm that sparked that desire to become teachers. When that locus of control is taken away from the individual and administrators constrain the individual and say "You can't do this" and "You can't do that; but we want to have output, we want to have professionalism, and we want to have good instruction," then there's a contradiction in terms.

Tucker: I want to bring a little history in. When my father got his Ph.D. during the Depression around 1932, the standard job for a beginning Ph.D. was 15 hours for $1200. Those people worked very hard and they had to get their papers out for tenure at top universities. Four-year state colleges tend to have twelve-hour teaching loads and they expect research for tenure. They generally don't have graders either. I think that at the community colleges there are some people who are very active and who formerly would have been teaching in the four-year colleges before there was this edict in which you had to have a Ph.D. to teach in college. Then there's another pool of teachers in two-year colleges who previously taught in high schools. High school teachers tend to work at their jobs for a certain number of years and, if they don't do anything really bad, they get tenure. So there are two types of tenure we are talking about; one that's earned from a high degree of professionalism and earned respect from the administrators, and there's another that's being in place--and I think it's very confusing to figure out what exactly tenure means at a two-year college and what level of respect one earns by getting that tenure.

Albers: A study that was done a few years ago by the Rocky Mountain Mathematics Consortium investigated professionalism of faculty across the country. It revealed regional variations, profound variations, stemming from their development modes. I'll just cite one area of the country. In the Northeast, you very often find two-year colleges existing as branches of the university system and in many cases that means that the attainment of tenure is done along fairly well-established lines. It doesn't seem to be all that easy for faculty to get tenure; the result is a lot of professional activity. Now, in other parts of the country the principal criteria for achieving tenure is not to "screw up" and to "teach okay" for a year or two. I'm not surprised that you get these variations based on these expectations.

Professionalism Versus Money

Leitzel: I don't think that we are seeing very much difference caused by amounts of money available to institutions. The honest thing to say is that the four-year institutions are strapped for money in the same way that the two-year institutions are. In Ohio, we think that the legislature is doing at least as well by the two-year colleges as it is by the four-year universities. Another thing that is lacking in the two-year structure is the mechanism of promotion for making visible that a person's work is appreciated and recognized. Faculty are motivated towards professional activity very strongly by the prospect of being moved from assistant professor to associate professor and ultimately to professor. I suspect this motivates faculty more than salary increases and reduced teaching loads. And that doesn't cost an institution very much!

Long: I think these things are widely variable depending on location and what may be true is different in Ohio and more different yet in Washington. With regard to your statement, Warren, about money and mandating professionalism, I'm not quite sure that I agree. I think that you need not reward people who do not produce. The idea of rank that Joan is espousing is used in certain community colleges, and I think effectively. So I think that various steps can be taken by administrations to establish an atmosphere that fosters professionalism. In other two-year colleges where these things are not done, you just don't get the professionalism.

Fusaro: My stomach churns a little bit when I hear all these comments about unprofessionalism. I just want to sound a cautionary note on self-flagellation, because earlier we said it was very important for faculties in two-year colleges to be viewed as professional if they wish to be recognized by funding agencies and to be listened to by four-year colleges and high schools. My impression was that two-year college teachers were more conscientious than those in the four-year colleges. I knew we four-year college teachers were more conscientious than those in the universities. My assumption and impression was that two-year college teachers were more conscientious, so, in a sense, more professional.

Ellis: Part of professionalism is getting each individual instructor to see that everything you do in your classroom is tied to the Board of Trustees somehow; and if you're going to get some money to get some colored chalk, there is a process to go through and you are in control of that process in some sense. If you want a piece of pink chalk in your tray, you can make that happen; and if you want a wide-screen TV in your classroom with a microcomputer hooked up to it, you can do it. It takes effort, but you are in control, and I think that's the part that's professionalism. You have to be in a situation where you think that what you do has an impact, and at our college most of our effort is focussed on the classroom. How are we going to make it better in there? Now, to make it better, we may have to get the heating system fixed (and this may be something that is specific to the West Valley Mathematics Department). The air conditioning did not work for ten years at West Valley College and every single year, at least once, the entire department went up to see the President. We put temperature gauges inside each room. What he said was, "Well, if you leave the doors open, surely the air conditioning won't work, so what you do is close all the doors." And then everybody passes out and that went on for ten years. After ten years--and think about this--ten years of "professional effort" on a trivial detail, we got it fixed. Now, when we start swapping lies in the mathematics department, part of our esprit de corps is based on that experience.

If we're going to make all of these changes, we have to affect each individual faculty member in each individual department. It's got to be at a local level with local control of your activities. That's the essence of professionalism in my view, and if you don't have the feeling that you can control them, you're never going to become a professional.

Gordon: Listening to most people talk here this morning, professionalism in the two-year colleges seems to be equated to math education or math instruction. I'd like to suggest that professionalism should also include mathematics. When we are teaching the same courses repeatedly, over and over again, where many of us and our colleagues may be lucky or unlucky, depending on our viewpoint, to get one section of calculus every few years, the rest of the mathematics that was learned becomes rusty from disuse--it just fades away.

Long: Quite apart from what we teach, we need to stay alive mathematically; and for our own mathematical well-being, we need to learn about fractals and all sorts of wonderful and exotic things that may have nothing to do, apparently, with what we teach. We got into the subject presumably because we like it; it turned us on somewhat; and we need to stay turned on or we're going to burn out.

Faculty Exchanges

Page: Suppose there was a duad of a senior institution with a community college, and each year, on a rotational basis, some two-year people would get a chance to go into the four-year college and partake of teaching new courses on a revitalization program. In exchange, some of the people in the four-year colleges would come to us and benefit from the exposure to two-year college student populations and teaching techniques? I would venture to bet that that would be very well-received, in terms of effectiveness; but I'm also willing to bet that the four-year people wouldn't want to do it.

Leitzel: We do bring to Main Campus (of Ohio State University) our two-year regional campus faculty on a rotating basis. It's hard for people to uproot often for a year and go out and do something in another location. In those communities that have both four-year colleges and two-year colleges, I think it would be excellent.

Rank and Tenure?

Warren: In terms of professionalism, I hear a frustration regarding the two-year institutions. Maybe we've lost the idea of the mission of the two-year colleges. We're teaching institutions, folks, and we're not research institutions. Now those are two extremes, but we have to look at the missions of the two institutions, and I think that the strength of two-year insitutions is that their faculty are _teaching_ faculty in _teaching_ institutions. I, for one, have always opposed putting professorial rank on two-year faculty. My reason for it is that I think it degrades the professorial system in the four-year institutions because it does require rigorous scholarship that is not necessarily required in two-year institutions. I think that it's unfortunate in New York City that they use the same criteria for promotion at two-year institutions that is used at four-year institutions. I can do much better if all my faculty are instructors than I can if they're professors, because they want the same breaks in terms of salary, sabbaticals, and so forth, but they don't want what it takes in terms of professorial strategies in four-year institutions; namely, that in six years you're up or out.

Albers: Two-year colleges have changed dramatically over the last 25 years. I'm not sure that their faculties can change as rapidly as the institutions have changed. The doors have been thrown wide open, and their foci are many now. Twenty-five years ago most of them had a single function and that was to prepare students for transfer to a four-year college. They had a strong liberal arts orientation. The faculty coming into these institutions by and large thought that they were going to be part of that enterprise--the liberal-arts transfer function. Meanwhile, these institutions broadened greatly. We now have great numbers of students coming to us with an occupational/technical goal in mind and that's producing some really wild tensions. It's produced what I've sometimes called a "faculty in limbo." We sometimes don't know who we are anymore.

Rodi: I find myself agreeing with a lot of what Bill Warren has said at the conference, but I got uncomfortable when you said that we may have lost sight of the mission of the two-year college--that we're teaching institutions. I hear too many administrators saying that sort of thing when they really mean "Now, look, all I want you to do is teach and that's it." Now here's what I think is the error in that thinking. A dean of Harvard recently wrote a piece for the CHRONICLE OF HIGHER EDUCATION. He talked about administrators being in a position where they are constantly drawing from their personal resources and never having much opportunity to replenish them. He was thinking of a university setting, where there are lots of opportunities to constantly replenish your personal resources. At a community college I think teaching is in a position where--like administration--and perhaps even worse than administration, you don't have an opportunity to replenish your personal resources. So, when I hear many administrators saying, "All we want you to do is teach," they're not paying any attention to how much water is left in the well. In all of our talk about professional development for community college faculty, I hear that same message that the dean of Harvard was giving--we need a chance to refill the bucket.

Part 7.

COORDINATING CURRICULUM CHANGES

COORDINATING CURRICULUM IN TWO-YEAR COLLEGES WITH BACCALAUREATE INSTITUTIONS

by

Betty Anne Case

and

Jerome A. Goldstein

SUMMARY

The success of transfer students from two-year colleges to baccalaureate institutions depends on a number of factors. Topics discussed here include policy and administrative matters, possible roles of professional societies, and baccalaureate institution responsibility after transfer. The main emphasis involves the important roles played by individual faculty members at both types of schools.

COORDINATING CURRICULUM IN TWO-YEAR COLLEGES WITH BACCALAUREATE INSTITUTIONS

Bettye Anne Case and Jerome A. Goldstein

1. Introductory Remarks

Coordination leading to successful student transfer experiences from the two-year college to the baccalaureate institution inherently involves policy and administrative as well as curricular matters. Many of these are beyond the ability of individual faculty to influence although we will discuss a few possibilities later. First, though, let us concentrate on coordination involving faculty primarily, distinguishing between formal coordination and local communication. Most of the *evidence* presented here to support suggestions is anecdotal, but hopefully there will be some suggestions which in context ring true to the reader. We do not expect to say much not already *known* to the reader, who will, we hope, be reminded of some useful ideas.

This conference and the joint MAA-AMATYC Panel on the Two-Year College Curriculum [1] are formal but non-mandatory coordination efforts. Carrying a greater aura of authority is *The Curriculum Guide for Baccalaureate Oriented Courses in Mathematics* [2] prepared by the Junior College Task Force affiliated with the Illinois Section of the MAA. It lists minimal course content and was "printed by authority of the State of Illinois" and carries some authority despite the absence of any enforcement intent. And in states with common course numbering in state schools, specifying a minimal core of common topics in courses similarly numbered increases further the mandatory nature of *coordination.*

2. Role of Professional Societies

Professional societies do not appropriately decree uniformity and cannot mandate change. However, it is within the scope of the professional societies to encourage the voluntary following of recommendations made by formal study projects and to disseminate these recommendations widely within their membership and to appropriate administrators. It is helpful when publications specifically address the two-year college situation; two recent M.A.A. publications notable in this regard are the C.U.P.M. panel reports *Minimal Mathematical Competencies for College Graduates* [3] and *Mathematics Appreciation Courses* [4].

In approaching this particular problem, curriculum coordination between two-year colleges and baccalaureate institutions, more than one professional society is involved as the people who need to work on the problem belong to at least three different national groups. Because of earlier teaching at the high school level or involvement in affiliated state two-year college groups, many two-year college faculty are involved in NCTM (which does not confine itself to K-12 interests). Other two-year college faculty are members of MAA, AMATYC or state (but not national) groups. One suggestion that has been made by a two-year college member of the two-year curriculum panel, Sheldon Gordon, is that the two parent organizations, MAA and AMATYC, hold joint national meeetings. We quote from a recent letter written by Gordon [5]. "The growth in AMATYC and the state-wide two-year college math associations is a positive step. It promotes a feeling of identity and unity among these faculty members and, as such, provides the opportunity to interact with mathematicians at other institutions. It is undeniably a healthy development. On the other hand, it carries with it some drawbacks as well. In particular, might it not create a further ghetto attitude, separating the two-year college mathematician from his counterparts in the four year schools? In turn, won't this institutionalize the feelings of inferiority? I suspect that they will, unless the organizations themselves realize the implications and address them. For instance, just as the AMS and the MAA hold their national meetings jointly, might not it be desirable to have AMATYC do the same?"

Perhaps some of the ideas in this quotation are controversial, but any generated controversy should lead to healthy communication. Professional societies, by the very act of holding meetings drawing a diverse population with common interests, act as a catalyst for the communication between faculty at different types of schools. When this occurs between neighboring two-year and baccalaureate faculty, automatically the case at state or section level meetings, concrete progress can be made toward curriculum coordination. Mathematics program heads in two-year colleges reported increased participation in conferences and the reading of journals by their faculties in 1975-1980. This bodes well for the opportunity for communication. (Cf. [6].)

And this communication on a local level, between two-year faculty and the baccalaureate faculty at institutions to which their students will transfer, is helpful even without strong formal efforts at coordination; it can lead to fine tuning so that for some students there is practically no transfer shock. Before we continue with suggestions concerning local communication, let us describe one professional society activity which is not widespread and which can be helpful in local curriculum coordination. Actually, D. R. Lichtenberg has coerced us into a two minute commercial for what the officers of the Florida Section of the MAA seem to think is the best Florida export since orange juice. (See [7]. These officers will happily supply more information.) Half day or Saturday joint meetings of high school, two-year college, baccalaureate institution faculties and other mathematics educators in geographic regions of radius under a hundred miles are organized under the auspices of the Section. These not only facilitate communication but often result in the two-year faculty gravitating to the central role in planning. In Florida this has encouraged somewhat more two-year faculty participation in the larger Section scene and considerably more respect for two-year college faculty by baccalaureate faculty, and vice versa. The real payoff is in improved articulation between high schools and colleges and between two-year colleges and baccalaureate programs in the region. Mathematical and pedagogical talks by participants from all schools, films and a formal articulation session for information sharing are valuable, but conversations over coffee or at dinner lead to a mutual respect which cannot exist outside personal acquaintance.

This Florida effort seems to be precisely the sort of thing, on perhaps a less formal scale, called for in the recent conference *New Goals for Mathematical Sciences Education*, sponsored by C.B.M.S.: "The conference recommends the establishment of a nationwide collection of local teacher support networks to link teachers with their colleagues at every level..."[8].

3. Informal Communication Channels

There are frequently no official channels for the exchange of curriculum and program information. Early information about prospective changes is invaluable. Our advice is: grab the telephone and make a serious effort to establish meaningful contact with sympathetic colleagues who could be helpful. Official channels may not be as useful as informal individual contacts. Mutual respect is likely to be established more quickly through informal contacts. Additionally, such contacts have another obvious advantage over contacts through official channels, namely, they can give indications of the directions and likelihood of possible changes.

Long before revised entry requirements for a program are approved by a university governing board, they are debated and refined by the faculty of that program, often as a matter of public record and rarely with ethical need for secrecy. Early explanation of what *may* happen from baccalaureate to two-year faculty colleagues could result in informal advice to two-year students which is invaluable. Baccalaureate faculty may hesitate to pass along the need for changes because they want to avoid looking dictatorial. But nevertheless they can ask two-year faculty about courses and texts, and they can offer information valuable to transfers or those preparing them to transfer. Such information may not be in print.

Two-year faculty can and must similarly instigate communication and must not be put off if the first person contacted, possibly a busy department chair, is not very responsive. *Some* faculty member at the baccalaureate institution does care about the mutual problem of transfer articulation, knows it cannot be successful without curriculum coordination, and is eager to work with two-year colleagues. We reaffirm our earlier advice: keep inquiring.

Once communication is established between individual faculty or administrators in neighboring mathematics departments, the obvious next step is a sharing of materials and methods. Syllabi, information provided to new instructors, flow charts showing sequences of topics within or between courses, and discussion of optional topics, will productively move in both directions. Baccalaureate faculty may be surprised to find sophisticated educational research or curriculum projects being carried out by two-year college faculty and can often benefit from their results.

One university, faced for the first time with teaching non-credit remedial courses and the further requirement that they contain certain specified *minimal skills* to be tested on a state-wide basis, made prompt (and thankful) use of the careful matrix of skills vs candidate texts made up by the local community college. In the other direction, a university statistics department teaching a two-semester service sequence, the first semester of which could be taken at the local community college, shared their exhaustive text analyses and the reasons for their rankings with the two-year faculty member, not a statistics specialist, teaching the course [9]. The two-year college adopted the same text and the instructor was careful not only to cover *Chapters 1 to n* but to begin the subject matter of *Chapter n + 1* and to review carefully concepts and techniques to be used in the sequel. The word quickly got out that students should be sure to get Statistics I at that two-year college because those who did made better grades in Statistics II [9].

We do not wish to imply that the same texts or methods must be adopted if articulation is to be successful. Another situation where there was communication led to different text selection but an interesting set of byproducts [9]. A two-year college faculty member who often taught the calculus sequence also taught an occasional night course at a nearby university as an adjunct instructor. Since he was known to some of the members of a committee revising the university calculus sequence, he was invited to sit on the committee. Despite some initial misgivings on both sides, he participated as a full committee member over several months of regular meetings and gave detailed and specific input. As it turned out the committee rejected for the university all those texts which omitted certain topics of the third semester of the sequence. The two-year participant felt the resulting choice was optimal at the university and voted for it but felt that a text rejected earlier had more examples and more careful explanation in its early chapters. When he explained the situation to his two-year colleagues, their decision was adoption of the text which had been rejected early at the university, reasoning that since there were relatively few students enrolling in Calculus III at their college, supplementing the text was a lesser problem than dealing with a less "user friendly" beginning. The real winners of this situation were the students transferring between these institutions.

4. Encouraging Administrative Responsiveness

There are limits to what can be achieved by individual communication or even department to department communication. Sooner or later there is the *administrative hurdle*. When it is a matter of the inability of two-year faculty to get an approval of a program change or such, patient explanation and a united front of the two-year department faculty will likely be successful.From the reports of former students, articulation success is at least anecdotally verifiable and administrators are susceptible to pressure to facilitate improvement. (We wish it were possible to loosen up travel money or lighten teaching loads for two-year faculty by just explaining the benefits!) Several MAA sections having articulation sessions at their meetings which are advertised under state auspices (or are publicized through notification of *academic deans* in addition to *department chairs*). These have been successful in expected ways as well as in unexpected ways. Namely, they have led to the availability of additional travel funds and, in the case of Illinois [10], have resulted in improved relationships with administrators.

It is generally felt that two-year colleges are relatively more responsive to community pressure and less tied to tradition than baccalaureate institutions. This very responsiveness points up the necessity for short-range communication. Interaction with nearby universities can affect two year college curriculum adaptation more than national studies For example, a current report on the status of discrete mathematics in two year colleges might conclude that few but large departments with depth in teaching resources offer discrete mathematics courses. But suppose that a two year college has a significant number of students per year intending to transfer into the computer science or mathematics curriculum at a nearby university teaching discrete mathematics to freshmen and sophomores and requiring it for all programming courses. If there are enough such students to run a small class, then *depth and resources* just may be found. While two-year faculty must keep aware of the big scene and national trends, change often cannot be made by two-year colleges in advance of implementation at the baccalaureate institution.

Students sometimes transfer, for academic or financial reasons, from a baccalaureate institution to a two-year college. Since these *reverse articulation* problems are not very common they probably do not need addressing but an anecdote about one may be worthwhile. A two-year college business calculus instructor had found that students who transferred from the nearby university between precalculus and calculus, although having studied from the same text, had covered one chapter less. Since these particular students were not academically strong anyway, this was a difficult problem. Discussing this with the associate chair of the mathematics department at the university, they agreed that the breaking point between the courses at the two-year college was more logical, and that the appropriate university committee would take it up. But, despite their personal agreement, there would be no change soon, although he commiserated with her [11]: "You people must sometimes feel like you're in bed with an elephant."

5. Caveats

Although certain caveats in communication between two-year and baccalaureate faculty may seem obvious though preachy, and others may be debatable, we cite a few for the sake of dialogue. It is best that baccalaureate program faculty not get involved or even state opinions about two-year college internal problems such as (1) department structure or power struggles, (2) conflicts between those two-year college faculty with allegiances to different professional organizations, or (3) discussion over which courses should be credit bearing at the two-year college. (This latter is not really a problem for the baccalaureate program standards because major requirements and admission entry levels can take care of it.) Baccalaureate faculty must not be didactic or proprietary about the absoluteness of the upper division status of a course. (It probably is, after all, taught to freshmen at some schools so such arguments will probably look foolish as well as selfish or bossy.) It is important not to state opinions which could possibly be construed as trying to dictate which *boundary* courses be left for the baccalaureate institution and certainly never be judgmental about *what the two-year college can teach "right"*. For example, if the first programming course for engineers at the baccalaureate institution uses Pascal and the two-year college still uses FORTRAN, that fact should be made known to the two-year faculty. Lack of action on that information does not mean disregard of advice or callousness to student needs. The two-year college computer system may not have a Pascal compiler, or students transferring in greater numbers to another baccalaureate institution may need FORTRAN. Two-year faculty should not assume Ph.D. research mathematicians do not care about quality teaching. In addition to cooling the chances of possible cooperative actions quickly, it also sometimes leads to failure of the two-year college to hire Ph.D. holders when they are available in a job applicant pool. Further they should not assume two-year faculty are not welcome at seminars, colloquia, and meeting sessions on mathematical topics. No one but the specialist understands all of the typical mathematical talk, and the two-year college faculty member whose interest is piqued by a topic is always welcome.

Of course neither should assume that there is a universal best way to do things. And neither should be too inventive or too quick to make changes which will affect articulation. In particular, credit courses should not be restructured so as to handicap students who transfer in the middle of course sequences. The very nature of the two-year college student body and the higher percentage of late bloomers there guarantees that at whatever point in mathematical training a student *can* transfer, some student *will*. Innovative restructuring leading to long sequences of courses needed intact will probably hurt some students so badly that its value is questionable.

6. Baccalaureate Department Responsibility after Transfer Occurs

This leads us to consider what can be done to ease two-year college baccalaureate institution articulation problems after the student transfers. Most of these are functions of the baccalaureate institution administration and faculty but one at least is a two-year college function: supplying information requested for transcript evaluation. The two-year college administrations are sometimes less than quick in responding to such requests for the statistically unusual student who insists on transferring from Michigan to Miami. If the request for information on a particular course reaches a two-year college mathematics department, an appropriate response would be the *immediate* dispatch of either a syllabus outline of topics with approximate time spent or, if the course was taught entirely from a standard text, a copy of the title page and table of contents with the covered portions marked. Often the administrative offices just send the same catalogue description that the baccalaureate institution had already found insufficient for evaluation.

The baccalaureate institution administration is responsible for transcript evaluation, advising, and placement procedures but will often respond to suggestions for improvements. Some flexibility to move students *back* (or forward) to another course at some early point is perceived in the sequel as ideal but rare. There is rarely any *grace period* for adjustment to the new college by transfers as is typically allowed for first time college students. Transfers with bad grades in their first semester are often not invited back. The only safeguard often is that they receive careful individually considered advice and that placement err if at all on the conservative side. While acknowledging that some students from two-year colleges just will not be able to make the transition, it is only humane to prevent instant and perhaps unnecessary washouts.

One incentive to voluntary curriculum coordination may be avoiding the big stick technique of the Florida legislature. Although it primarily benefits students the articulation agreement legislated between the public two-year and baccalaureate schools in that state can slow necessary change and stifle incentive to innovate.

A minimal fall-back position for a baccalaureate faculty member or department desiring an improved transfer success rate is to be very careful that individual advisees are properly placed. Even with the best of intentions and with information about previous courses, very hard questions remain. What does one do about the mid-sequence student with Calculus I credit but neither the integration applications or techniques material that the baccalaureate school teaches in Calculus I? What about the student transferring in with a computational linear algebra course when the local course is a heavily proof oriented preparation for modern algebra and advanced calculus? In some of these cases, auditing the earlier course is the best advice for an average student. However, considerable time and explanation must be spent with students before they can accept this recommendation and not regard it as a *put down* of their earlier training.

Individual faculty can, through the classroom, improve the transfer success rate. Four specifics which do not take a great deal of effort or involve actual course slowdown and which benefit all students are: 1) Giving transfer students and others desiring it a copy of the preceding course syllabus. (This will prompt some "I never had..." comments from adequately prepared students but will warn those deliberately or accidentally overplaced.) 2) A *review* assignment with considerable specificity can be made for the first weekend of the course and can include minimal aspects of the prerequisite material. 3) A *makeup* lecture or two or an arrangement with the Math Lab or tutoring service to provide this may be possible. (And, somehow, the students needing this early intensive work must be identified and persuaded to take advantage of these first three assistance devices.) 4) It must be remembered that having transfers requires extreme care to re-teach earlier concepts when they are needed, especially such concepts as are known to be not well understood.

7. Conclusions

While the baccalaureate faculty must bear much responsibility for realistic placement and assistance in adjusting, the two-year faculty must be certain that the student passed with a "C" is ready for transfer at *some* level. It does not matter greatly in the long run whether the student gets through course p or p + 1 at the two-year college, but there is a problem if each course is less demanding than its counterpart or is inadequate preparation for the next logical course placement at the baccalaureate institution. The often repeated motto of many open admissions school faculty: "Our school may be easy to enter, but our students are worth something if they graduate", should be borne in mind with respect to the courses in the baccalaureate track mathematics offerings of the two-year college.

The desire by both institutions that students be successful is the strongest pull toward working out articulation problems, and that pull may be strongest on the two-year college. Tenacity on the part of the two-year faculty to obtain the information needed to be sure their curriculum is consistent with successful articulation will produce contacts with baccalaureate faculty who can provide that information and who can offer the two-year faculty whatever association is desired in the scientific and pedagogical programs of the department. Baccalaureate departments which are the most cooperative with two-year faculty may discover the fringe benefit that the better students are encouraged to go on to their institutions, too.

Professional societies must meet their responsibilities concerning important issues in mathematical education in the first two years, namely, curriculum, teaching and coordination. This conference and the joint panel on the Two-Year College Curriculum constitute a good start on the solution of this problem, which one can argue has received too little attention in recent years [12].

While Lipman Bers was President of the AMS, he encouraged research departments to allow all possible access to libraries, offices, facilities, speakers, and so on to mathematicians at nearby non-research institutions. (Cf. [13].) We would urge that this suggestion be interpreted and implemented in the broadest practical sense with respect to both the mathematical and curricular concerns of two-year college faculty.

NOTES AND REFERENCES

[1] In 1983 the Committee on the Undergraduate Program in Mathematics and the Committee on Two-Year Colleges of the MAA and AMATYC formed this Panel. Chaired by Ronald M. Davis, it is studying global curricular questions and can be viewed as a two-year college analogue of the CUPM Panel chaired by Alan Tucker which produced *Recommendations for a General Mathematical Sciences Program*, MAA, 1981.

[2] Junior College Task Force of Illinois Section, M.A.A., *Curriculum Guide for Baccalaureate Oriented Courses in Mathematics*, April 1982.

[3] D. W. Bushaw, Panel Chairman, Minimal mathematical competencies for college graduates, Amer. Math. Monthly 89 (1982), 266-272; available from the M.A.A. in reprint form.

[4] J. A. Goldstein, Panel Chairman, Mathematics appreciation courses, Amer. Math. Monthly 90(1983), 44-51 and C11-20; available from the M.A.A. in reprint form.

[5] Letter from S. P. Gordon to R. M. Davis, dated 21 October, 1983.

[6] J. T. Fey, D. J. Albers and W. H. Fleming, *Undergraduate Mathematical Sciences in Universities, Four-Year Colleges, and Two-Year Colleges*, 1980-81, C.B.M.S. (1981), 110 pp.

[7] B. A. Case and D. R. Lichtenberg, Regional meetings within the Florida Section, Focus, to appear.

[8] *New Goals for Mathematical Sciences Education*, The report of a conference sponsored by the CBMS, Notices Amer. Math. Soc. 31 (1984), 260-270.

[9] These experiences involved Tallahassee Community College and the relationship with the Department of Mathematics and Computer Science and the Department of Statistics of Florida State University, 1977-1984.

[10] See [2].

[11] R. McWilliams of Florida State University to B. A. Case of Tallahassee Community College in 1977.

[12] See [1] and *Two-Year Colleges and Basic Mathematics*, Reports of the CUPM Panel on Two-Year Colleges, in A Compendium of CUPM Recommendations, Vol. I, MAA, Washington, D. C. (1969, 1971), 203-367. Included is: A transfer curriculum in mathematics for two-year colleges (1969), 205-255.

[13] L. Bers, Letter to the Editor in Notices Amer. Math. Soc. 23 (1976), 216.

(DISCUSSION BEGINS ON P. 471.)

SOME REFLECTIONS ON THE INTERACTION OF MATHEMATICS PROGRAMS AT TWO- AND FOUR-YEAR COLLEGES

by

Stephen Rodi

SUMMARY

Increasing numbers of students will transfer from two to four year colleges. In light of this, two and four year mathematics faculties need to sharpen their understanding of each other and to plan improved interaction.

For the most part, faculties at two and four year schools live in separate worlds. Part of the reason is mutual uncomfortableness rooted in mutual stereotyping: the TYCer is suspected of being standardless and/or of being a pseudo mathematician; the university professor is viewed as aloof and ineffective with undergraduate students.

Both groups need to come to recognize that the other has unique and important contributions to make in a dialogue. The TYCer needs to visit the four year campus regularly--and teach and work there from time to time--to re-experience the environment his transfer students will encounter. The university faculty need to recognize--and learn from--the outstanding teaching that generally takes place at two year colleges. The groups need to work together on a variety of projects and programs that can benefit each of them and students. In particular, they need to interact in discussions of mutual curriculum and standards, of entry level preparation of incoming students, and of details of mandated state curricula, where such exist.

Mutual respect is the best foundation for future cooperation. Interaction, particularly at the local level, is the best way to promote respect. The principal professional mathematics organizations have a special responsibility to promote this interaction, where necessary changing their policies and emphases and encouraging attitudinal change in their members. In short, for a relationship to work, the partners need to see each other as equals.

SOME REFLECTIONS ON THE INTERACTION OF MATHEMATICS PROGRAMS AT TWO AND FOUR YEAR COLLEGES

Stephen B. Rodi

Talk Net is a syndicated nationwide phone-in radio show with early evening host Bruce Williams. Williams dispenses wisdom on the best swimming pool liners and on the worst life insurance policies. His fans love him.

On the night I drove to my office to begin sketching out these reflections, one young man called Williams to discuss the merits of staying at the local community college for a year or two before transferring to the four year college. Williams' reply: you certainly will save some dollars that way, a plus in anyone's ledger.

In his own unvarnished way, that night Williams gave us the principal reason two and four year mathematics programs need to be concerned about one another. For the rest of the century, and likely beyond, economics will motivate (and frequently force) large numbers of students to start out at the community college and end up at the four year school.

I know that the 1980-1981 CBMS survey committee report (1,Figure 4.2) showed that college transfer enrollments as a percentage of fulltime enrollments at community colleges had dropped below occupational/technical enrollments for the first time. But I sense that phenomenon may be reversing again. Or, even if these percentages do not reverse, the gross number of students at two year schools who transfer to four year schools certainly will grow steadily.

I know this is the case at my school where each semester we have over 800 students in calculus and differential equations but fewer than 200 in arithmetic. I sense it is a growing phenomenon in my state where, for example, increasing numbers of community colleges are concerned about topics like the common core engineering curriculum for freshmen and sophomores at publicly supported colleges. (By the way, "common core" does not mean all colleges must offer the same courses. More on that below.) When a state like Florida has a 357,993 headcount in community colleges (December, 1982, figures), a transfer rate of but 20% feeds 70,000 students into a university system whose headcount at that time was only 135,072. Finally, at the most recent meeting of the American Association of Community and Junior Colleges (AACJC), as reported in the April 14, 1984, Chronicle of Higher Education (p. 22), the dean of instruction at Virginia's Piedmont Community College predicted that the increasing cost of higher education would lead to larger enrollments of middle-class, traditional, college-age students in two year college transfer programs.

Mutual Perceptions

It is hard to know exactly what perceptions two and four year mathematics faculty currently have of one another. But, if they are going to be sharing so many students, they ought to try to find out.

On the one hand, in the 1980-1981 CBMS survey (1, p. 111) more than half the administrators in two year college mathematics programs reported no problems with coordination with four year colleges. On the other hand, in this same report (p. 92), these same departments reported a low level of consultation with four year programs: less than once a year for 42%, only yearly for another 35%, and more than once a year for only 23%.

My own suspicion is that these last figures tell the real story: life for the most part in separate worlds. I suspect there are two causes for the infrequent interaction. One is the natural human inertia and organizational problems in getting one's own department together without even considering coordination with the school across town. But a second reason is a fundamental uncomfortableness between mathematics faculties at two and four year schools, an uncomfortableness that has existed for most of the past two decades and only now may be diminishing slightly as these faculty interact more on national committees and at each other's national and regional meetings.

As in many such cases, some partly factual, partly stereotypic attitudes may lie at the root of the uncomfortableness. The university faculty member isn't quite sure the two year faculty member _really_ is a mathematician and for that reason is reluctant _fully_ to welcome the TYCer into the informal camaraderie of the mathematical fraternity and into the older, mainline mathematical organizations. Two year faculty members, for their part, perceive too much casual disregard for undergraduate instruction at large four year schools where massive lecture sections and the faculty's pursuit of publication isolate the student and result in many student failures that should not occur.

I do not think we should pretend these mutual perceptions are not there. They are. And they are too often grounded in fact. But, in an era when cooperation will become more critical, we need--as a first step toward improved interaction--to work at modifying these perceptions.

Learning From Each Other

An important first step in this process of change is for each group to recognize that it has something to learn from the other. Each can look to the other for improvement.

Too frequently two year college faculty are not good enough mathematicians. Their mathematical education has not been broad enough. They need to be willing to expand it, particularly in a decade that will bring more probability, statistics, discrete mathematics, and programming to their courses, even at introductory levels. And their view of what they do as teachers--particularly with transfer students--can be too narrow. "Feeling good" about factoring is not an adequate goal. Too many TYC faculty are influenced inappropriately by training programs based in colleges of education which confuse ideas with feelings and seem to make the touchstone of the successful two year college a student exiting with a "warm glow" about a "good experience."

In short, these TYC faculty, like some of their colleagues in secondary schools, need to raise their personal and departmental standards, particularly as regards transfer students. Such action will have a positive effect on the confidence and respect accorded by their four year colleagues.

Colleagues at four year schools can be helpful in this process. For example, an imaginative program at The University of Texas at Austin invites mathematics faculty from across the state to spend a "semi-sabbatical" in Austin teaching at U.T., participating in seminars, and refurbishing research skills. As it stands, that program is skewed a bit toward other four year faculty. But it is a good, solid model of a program that also could be oriented for visiting two year faculty, even if only in the summer. The TYCer needs to get close from time to time to the environment the transfer student will end up in and needs to have some intellectual challenge similar to that his transfer students will face.

But four year faculty have to recognize that two year faculty have special skills to offer, too, in this dialogue of recognizing each other's strong points. It may well be that much of the best teaching in the U.S. today at any level takes place in the two year colleges. Four year faculty ought to be more aware of this and more eager to tap this resource.

All sorts of possibilities exist. Graduate students could spend an intern semester teaching with two year faculty. An experienced, effective two year faculty member could be invited to the four year school to teach undergraduate courses as models for graduate students and others. The four year department might invite their neighboring two year colleagues to participate in discussions on curriculum or classroom organization or teaching styles and effectiveness or textbook choice for undergraduate courses.

Wouldn't it be lovely (apologies to Henry Higgins) if graduate programs could take at least as much interest in the teaching skills of their students as in their research ability? After all, not only will their graduates be entrusted with the education of many generations of young people--no small societal responsibility--but faculty salaries are paid by parents and taxpayers who expect and deserve good instruction for their dollar--no small obligation in justice for the payee. Two year colleges could contribute to a heightened awareness of the importance of teaching.

Other Interactions

The needs of both four and two year college faculty for renewal and expansion of viewpoint and skills is one of the principal recommendations in the CBMS report "New Goals for Mathematical Sciences Education" (2, p. 20,21), the results of a conference held November 13-15, 1983. The suggestion here is that in some areas these two groups should look to each other for renewal. Here are some additional specific suggestions for interaction and cooperation where the beneficiaries are students and society.

When a state like Michigan (The Chronicle of Higher Education, April 18, 1984, p. 14) calls for colleges to pool resources, neighboring faculty at two and four year schools should be comfortable with each other to consult and see how coordination might help each other. When the Rockefeller Foundation reports (same Chronicle, same page) that lack of high school mathematics is a major hinderance to women and minorities in science, local community colleges and four year schools should be active, natural allies in attacking the problem.

I participate in two programs of the last kind, neither perfect, but both steps in the right direction. One is called TAME, Texas Alliance for Minorities in Engineering, largely funded by industry. In my area, it is not unusual for a graduating minority high school senior to spend the summer on scholarship at our community college in mathematics and physics classes before moving on to the university.

The second program developed five years ago when, for a variety of reasons, the mathematics department at The University of Texas at Austin decided to stop offering college algebra except in the summer term. Nonetheless, they expected a certain number of incoming freshmen not to score high enough on their placement examinations for the usual first level course (frequently, Business Calculus).

The mathematics chairperson at The University of Texas approached us at the community college. What developed was a Monday/Wednesday night college algebra class supervised and taught by us for credit at our college but offered in the U.T. mathematics building for mostly U.T. students. The course enrolls 400 total in fall and spring. (An interesting side note on this course. Four years ago we had over 500 requests from incoming university freshmen for the 200 seats available in the fall. As the university has significantly raised its entrance requirements, as well as adjusted downward somewhat the placement score acceptable in mathematics for first level courses, that number has dropped to about 200.)

In some ways, the most instructive part of this last example is the high spirit of cooperation that has developed among us, the U.T. mathematics department, the undergraduate advisors in the U.T. College of Business, the staff of the U.T. Measurement and Evaluation Center, and the summer orientation staff in the Dean of Students office at U.T. By getting to know one another over the phone and in person, each side has been able to make small adjustments that makes the program work for students. On our part, we have established a special mail-in registration inquiry that U.T. advisors can give their prospective freshmen. We have tailored the course to 15 weeks (rather than 16) to avoid conflict with their examination period. On the other hand, various colleges at U.T. have been cooperative in giving exemptions to the usual requirements of fulltime registration, if the student is enrolled with us for this special program at an advisor's suggestion. And the U.T. mathematics department through its chairperson has been most helpful in greasing the wheels of approval at various decision levels within his institution.

What Is A Freshman Course?

My personal experience with our college algebra classes on a university campus always makes me think of a particular area where I feel four year schools have something to learn from two year colleges. One result of the post-Sputnik, new math era was an attempt to put more preparatory mathematics in high schools so that almost all stuents entering the university and registering for mathematics would go into a calculus class. In many ways, this was in contrast to the practice of the 1940's and 1950's--when most of us went to college!--wherein the normal expectation was that a freshman college student would take a full year of mathematics (college algebra, trigonometry, a thorough analytical geometry course including rotations, and maybe even a little spherical trigonometry) before beginning calculus.

In my mind, the curriculum revision of the past 20 years clearly has failed. Our high schools are graduating seniors distinctly less well prepared in mathematics. Nonetheless, there remains at most universities the expectation and assumption that calculus is the normal freshman course, that students who do not begin calculus in the first semester of their freshman year are "out of sequence" or "doing catch up work."

The pressures for calculus as a freshman course frequently come from outside the mathematics department (e.g., engineering and physics) where curricula are constructed on that assumption. But it may be time now to question the reasonableness of that assumption--or even the reasonableness of a four year degree plan--for the American educational system, at least until such time (should it ever come) when high schools will graduate more students truly prepared for college work. Two year colleges generally have been more realistic about the preparation level of students and could make a special contribution in a dialogue about entry level mathematics courses.

Common Curricula

The specific examples of cooperation mentioned above are valuable mostly at the local level. There is a broader area of cooperation of great importance. This has to do with establishing common statewide or regional curricula.

Robert E. Shepack is the president of El Paso County Community College. In a long article about Hispanic students, he is quoted in *The Chronicle of Higher Education*, March 21, 1984, as follows on the subject of transferability and program coordination between two and four year schools: "The problem is not a community college problem; it is a university problem. Universities have to develop enveloping curriculums so they can recognize what we do but still get what they want. And they have to be less imposing... They are the ones who refuse to change."

In El Paso's case, forty parallel programs have been developed in cooperation with the university; but, says Shepack, "only a few places in the country have well-thought-out programs that follow through a matriculation."

Shepack is commenting on what I think should become the single most important point of articulation between two and four year colleges for the rest of this century. Four year schools have to be willing to incorporate two year schools in the degree planning process so that the community college can organize its own curriculum to make the student's path straight and efficient. This interactive process requires confidence on the part of the university that transfer student indeed will receive a comparable education at the two year feeder school and requires flexibility where necessary at the two year level to make adjustments to earn this confidence from their university colleagues.

On one level, these sorts of common curricula can be imposed by the state whose legislatures have a powerful interest in not duplicating educational appropriations. Florida has an articulation agreement mandated by the legislature and supervised by a Deputy Commissioner of Education. Texas has developed about eighteen "core curricula" which assure intrastate transferability whenever public colleges offer the same course on the core list. In the 1980-1981 CBMS survey (1,p.92 thirty-eight percent of the two year colleges reporting indicated official state-wide coordination of two year mathematics courses with those of four year institutions.

It might be very instructive as part of the next CBMS survey, or as a follow-up from this Sloan conference, to get a great deal more detailed information on how the various states are handling this coordination and (where successful plans have been developed) to publicize the techniques. I have in mind something as thorough as the fifty-state survey recently completed by the Education Commission of the States on state response to the national crisis in precollege mathematics and science (3).

Frankly, however, I agree with Shepack that the biggest obstacle to the formation of coordinated curricula remains with the four year schools. No amount of state mandating will be truly effective as long as four year schools are reluctant to participate or are suspicious that their participation will in some way dilute their product. Too few large university mathematics faculties are like the University of Illinois at Champaign which took the initiative years ago to get involved with two year colleges and in the process helped establish one of the strongest statewide two year college mathematics associations in the U.S.

Promoting Mutual Respect

But Illinois--and some others, like Michigan and Florida--are showing us the right path: interaction at the local level. National groups like the American Mathematical Association of Two Year Colleges (AMATYC) and the Mathematical Association of America (MAA) have to promote both nationally and statewide as much collegiality as possible. MAA particularly, which through the years and even now sends ambivalent signals about its involvement with mathematics at the two year college, should use its prestige and resources to promote this regional cooperation by appropriate policy decisions and direction at the highest level.

Knowledge through association will erase misconception and become the foundation of confidence and cooperation. The closer to home the elbow rubbing, the better.

Much of what I have written in these reflections has to do with attitudinal changes on both sides that should facilitate cooperation between two and four year mathematics faculties. The March 28, 1984, *Chronicle of Higher Education* ran a provocative article about similar collaboration between college faculty and secondary school teachers. It was titled "Equal Status for Schoolteachers, Professors Called Key to Successful Collaboration." *Mutatis mutandis*, the article provides a good source of reflection on the relation between two and four year mathematics faculties. Let me close with some excerpts:

(a) "unless you enter a relationship believing you are working with persons who are equal to you, a project has no probability of success";

(b) "professors are not trained as teachers; they regard instruction as an obligation"; "when it comes to instruction, schoolteachers may actually be somewhat superior to faculty members...perhaps they could work with new faculty members and teach them some skills";

(c) "a school-college collaborative must develop a common understanding about what it is about";

(d) "there is a need to avoid the condescension that has marked the attitude of many college faculty members working with their counterparts at the secondary school level";

(e) "estabish an analogue of the county medical society for those teaching in each discipline... doctors meet monthly in county medical societies and take the primary responsibility together for the quality of practice of medicine in their area and for keeping each other up to date in the field...societies of...mathematics teachers should also meet monthly and take responsibility for the quality of teaching in their disciplines in their locale and keep each other up to date".

References

1. J. Fey, D.J. Albers, and W.H. Fleming. Undergraduate Mathematical Sciences in Universities, Four-Year Colleges, and Two-Year Colleges, 1980-1981, Washington, D.C.; Conference Board of the Mathematical Sciences.

2. "New Goals For Mathematical Sciences Education," Report of a Conference Sponsored by the Conference Board of the Mathematical Sciences, Washington, D.C., November 13-15, 1983.

3. "A 50-State Survey of Initiatives in Science, Mathematics and Computer Education," No. SM-83-1, Denver, Colorado, September, 1983: Education Commission of the States.

(DISCUSSION BEGINS ON P. 471.)

MATHEMATICS PROGRAMS IN HIGH SCHOOLS AND TWO-YEAR COLLEGES

by

Ross Taylor

MATHEMATICS PROGRAMS IN HIGH SCHOOLS AND TWO-YEAR COLLEGES

by Ross Taylor

SUMMARY

Today school mathematics is receiving greater attention than at any time since the flurry of activities following the launching of Sputnik in 1957. A Nation At Risk and many other national reports have called for an increased emphasis on mathematics. As we move into the information age, a person's mathematical knowledge will be a significant factor in determining whether the person will be a "have" or a "have not" in the new society.

Until now the secondary school mathematics program has tended to concentrate on two areas: preparation for four-year colleges, and basic skills instruction. The flexible curriculum to meet the diverse needs of all students called for in the National Council of Teachers of Mathematics Agenda for Action has yet to materialize.

Achievement of seventeen-year-olds has been stable in the 1973, 1978 and 1982 national assessments of mathematics. Students tend to be relatively strong in whole number computation, but they are weak in higher level mathematical skills. In many areas of the country there is a shortage of qualified mathematics teachers.

In the future we can anticipate greater curriculum emphasis in discrete mathematics, statistics and probability, and computer science. We can expect to see gradual infusion of recommended changes into textbooks and hence into instruction. The impact of computers and calculators on instruction will likely be uneven from school to school. Standardized tests will tend to follow the trends, rather than lead them. However, locally developed criterion referenced tests and College Board tests could be trend setters. In the absence of federal initiatives, curriculum changes will tend to be more local in nature than they were in the wake of Sputnik.

Two-year colleges should take a proactive role and work directly with their feeder high schools to help them give their students better preparation for post-secondary education.

MATHEMATICS PROGRAMS IN HIGH SCHOOLS AND TWO YEAR COLLEGES

by Ross Taylor

High School Mathematics Programs

Recommendation 6 of An Agenda for Action: Recommendations for School Mathematics of the 1980s states that:

> More mathematics study be required for all students and a flexible mathematics curriculum with a greater range of options be designed to accommodate the diverse needs of the student population.

The Agenda goes on to recommend that at least three years of mathematics should be required in grades 9 through 12. The report of the National Commission on Excellence in Education: A Nation at Risk also recommends that at least three years of mathematics and a half year of computer science be required in grades 9 through 12. If these recommendations were implemented, the result would be a significant change in the mathematics requirements in high schools throughout the country. Today, many of the nation's schools require only one year of mathematics in grades 9 through 12; a lesser number require two years of mathematics and still fewer require three years.

Up until this time, the high school mathematics curriculum has been essentially a two-track program. The precollege program has focused on precalculus mathematics including programs in elementary algebra, geometry, advanced algebra and precalculus. Typically, trigonometry has been included in combination with advanced algebra, precalculus, or both. The program for the noncollege bound has consisted of a year of general mathematics at ninth grade followed up in some cases with a year or more of consumer or vocational oriented basic mathematics. The Position Paper on Basic Mathematical Skills of the National Council of Supervsors of Mathematics (NCSM) states that all students need a mathematics program that includes more than computational skill. In particular, the NCSM Position Paper established that all students should aquire skill in the following ten vital areas:

1. Problem Solving
2. Applying Mathematics to Everyday Situations
3. Alertness to Reasonableness of Results
4. Estimation and Approximation

5. Appropriate Computational Skills
6. Geometry
7. Measurement
8. Reading, Interpreting, and Constructing Tables, Charts and Graphs
9. Using Mathematics to Predict
10. Computer Literacy

The NCSM Position Paper also points out that employment and educational opportunities will continue to increase as mathematical skills continue to grow beyond the ten basic skill areas.

As we move from an industrial society to an information society, one's opportunities will be determined to an increasing extent by one's ability to function with technology. In addition to communication skills, one's mathematical background will be a significant determiner of whether one will be a "have" or a "have not" in future society.

The transition from a two track (or more accurately a one plus a fraction track) mathematics curriculum to a flexible mathematics curriculum that meets the diverse needs of all students is neither swift nor easy. Mathematics teachers are products of college preparatory precollege mathematics. Most of them received their undergraduate education in four-year colleges and universities. Morever, most school administrators and counselors also received their undergraduate training in four-year institutions. As a result, their education and experience does not orient them to the two-year colleges. Moreover, the attention of the schools tends to focus on the ends of the academic spectrum. From the middle seventies on, the minimum competency programs spurred by the back to the basics forces have tended to focus on students at the lower end. In the eighties, the code word has been excellence with a resulting focus on gifted and talented students. In the process, the large group of students in the academic middle have not received the attention they deserve. To a large extent, this is the group of students that is served by community colleges. In the absence of more appropriate mathematics programs and better counseling for those students, the two-year colleges have had to dissipate a large proportion of their resources on remedial programs that address mathematical competencies that students should have acquired in high school, in junior high school, or even in elementary school.

Achievement Trends

As two-year colleges plan their mathematics programs for the rest of the eighties and on into the nineties, the planners need to be aware of current trends in elementary and secondary schools that will affect them in the future. Achievement of seventeen-year-olds has remained relatively constant in the three national mathematical assessments in 1973, 1978 and 1982. Achievement of thirteen-year-olds increased from 1978 to 1982. Achievement of Blacks and Hispanics is far below that of White students at all three age levels. However, the gaps have been decreasing for nine-year-olds since 1973 and for thirteen-year-olds since 1978. The infusion of federal Title I/Chapter I funds into compensatory programs in urban elementary schools is one of several possible explanations for these trends. In the future, the increases in achievement for all students and the decreases in the achievement gaps between minority and majority students may show up at the seventeen-year-old level. Whether or not these trends will continue as the students move through the secondary schools remains to be seen. There is no federal effort at the senior high school comparable to the Chapter I program which spends over three billion dollars annually in compensatory programs in reading and mathematics concentrated at the elementary level. In any case, students tend to achieve well on lower level skills such as whole number computation and they tend to have difficulty with higher level skills such as problem solving. No immediate turnaround in this pattern appears to be likely. Mathematics educators are particularly concerned about the low level of achievement in higher level thinking skills because these are the skills that students will need in order to be effective contributors in future society.

The long gradual downward slide in Scholastic Aptitude Test (SAT) scores appears to have bottomed out.

Public support for mathematics instruction is high. For example, in the 1983 Gallup Poll on Education mathematics was rated as the most important of all of the subjects both for college-bound and noncollege-bound students. The school effectiveness movement has taken hold with school boards and top school administrators. The result has been an emphasis on practices that raise achievement, particularly in the areas of reading in mathematics. The flames of public interest fanned by national reports such as A Nation at Risk have produced pressures at that local and state levels to increase high school mathematics requirements and upgrade the mathematics curriculum. In 1983 the National Commission on Excellence in Education reported that the nation's schools were being overwhelmed in a "rising tide of mediocrity." A year later, the U.S. Department of Education issued a report entitled The Nation Responds that claimed that the report of the National Commission on Excellence had produced "a tidal wave of school reform." The Department of Education 1984 report cited 275 state level task forces that had worked during 1983-84 to improve education. A total of 48 states are considering stiffer high school graduation requirements and 35 already have approved changes. In the present national environment, the stiffer requirements will probably affect mathematics as much or more than any other discipline.

The Mathematics Teacher Shortage

A shortage of qualified mathematics teachers continues to be a problem in many areas of the country. However, the shortage appears to be spotty. For example, high schools in affluent areas are generally well staffed. However, there are shortages in many rural and urban school systems, particularly in the junior high and middle schools. Reports from schools of education indicate that undergraduate students have heard about the shortage of mathematics teachers and, as a result, many more are coming through the pipeline. For example, since 1980 the number of undergraduates in mathematics education at the University of Minnesota has increased from under 40 to over 100. In addition, a number of teachers from other fields have entered graduate retread programs to prepare them to teach mathematics. Unfortunately, some areas of the country have addressed the issue by relaxing certification standards so that warm bodies could be found to staff the classrooms.

Sputnik and A Nation at Risk

Perhaps A Nation at Risk, the report of the Commission on Excellence in Education, will identify 1983 as a major reference point in educational reform, just as the launching of Sputnik identified 1957 as a benchmark year. In each case, international competition was a major factor that gave impetus to the reform. In 1957 the Russians were the competitors and the area of competition was aerospace and military technology. In 1984 the major competitors are the Japanese and the area of competition is economic with concentration on high technology industries. At this time we are seeing an emphasis on improving education that appears to be at least as intense as the efforts that followed the launching of Sputnik in 1957. Sputnik was accompanied by massive national curriculum development efforts in science and mathematics, funded primarily by the National Science Foundation (NSF). In addition, NSF funded numerous teacher education programs for upgrading the mathematics and science knowledge of teachers. The NSF summer and academic year programs for teachers tended to be national in scope with teachers from all parts of the country attending institutes at various colleges and universities.

The patterns for improvement today are much different than they were in the wake of Sputnik. Initiatives from the federal government have been slow in coming, so responsibility has fallen more at the state and local levels. At this time there are no massive federally funded curriculum development projects such as the School Mathematics Study Group (SMSG) or the University of Illinois Committee on School Mathematics (UICSM) of the late fifties and the sixties. For similar activities to take place today, a dramatic shift in emphasis in Washington or major private funding would be required. Similarly, today staff development efforts are much more local in nature than they were following Sputnik. In the absence of federal initiatives, states are allocating resources for staff development. For example this year Minnesota has allocated over half a million dollars for teacher education programs in the areas of science, mathematics and social science; in 1984 Florida allocated 9.4 million dollars for staff development and summer programs in mathematics and science.

The one area where today's efforts are national in scope is that of holding conferences. One leader in mathematics education indicated to me recently that if the problems in mathematics education could be solved by conferences, they certainly would all be solved by now. Of course, each conference produces a report and, in addition, there are individually commissioned reports not related to specific conferences. This situation caused a mathematics educator to remark that right now we are in less danger of being overwhelmed by a rising tide of mediocrity than we are of being overwhelmed in a rising tide of reports on education.

Future Curriculum Trends

Of the many recent reports on education, two that could have significant impact on mathematics education are the Conference Board of the Mathematical Sciences report to the National Science Board (NSB) Commission on Precollege Education in Mathematics, Science and Technology and the report of project EQuality of the College Board. Both of these reports were developed prior to the publication of *A Nation at Risk*. The two reports are general in nature, leaving the details to be filled in at a later time, yet they serve as an indication of future trends.

The report of the Conference Board is entitled "The Mathematical Sciences Curriculum K-12: What is still fundamental and What is Not." Here is a summary of the recommendations with regard to the secondary school mathematics curriculum.

We recommend:

> That the traditional component of the secondary school curriculum be streamlined to make room for important new topics. The content, emphasis and approaches of courses in algebra, geometry, precalculus, and trigonometry need to be reexamined in light of new computer technologies.
>
> That discrete mathematics, statistics and probability, and computer science now be regarded as "fundamental" and that appropriate topics and techniques from these subjects be introduced into the curriculum. Computer programming should be included at least for college-bound students.

The report of project EQuality entitled "Academic Preparation for College: What Students Need to Know and Be Able to Do." states:

> Students going to college need mathematical skills beyond the elementary ones. They need a knowledge of computing to deal with the new age of computers and information systems. They need a knowledge of algebra, geometry, and functions to major in a wide range of fields from archaeology to zoology. They need a knowledge of statistics for such fields as business, psychology, and economics.
>
> More extensive knowledge and skills, including preparation for calculus, will be needed by college entrants who expect to take advanced mathematics courses or to major in such fields as engineering, economics, premedicine, computer science, or the natural sciences.

Today mathematics instruction in the schools tends to be influenced primarily by the textbooks. Publishers are quick to pick up on new trends while at the same time being cautious not to publish radically different texts before the consumers are ready to purchase them. If present trends continue, in coming years we will see a gradual infusion of the recommended changes into the texts and consequently into instruction.

The Impact of Computer and Calculators

The change needed as a result of new computer and calculator technology causes a particular dilemma in mathematics education today. The technology is changing so rapidly that it is difficult to know what students will need in the future. For example, how much programming will students need, and in what languges? Will changes in hardware and software make programming as we know it today largely obsolete within a few years? Inexpensive hand-held calculators have been available since the middle seventies, yet the impact of calculators on the mathematics curriculum has been relatively minimal.

Use of computer timesharing in mathematics instruction appeared to flatten out at a relatively low level during the early seventies. The microcomputer boom in the early eighties has caused a spurt in the use of computers in all aspects of education. Yet today mathematics teachers use computers primarily for drill and practice in low level skills and for teaching courses in programming. The recommendation of the National Council of Teachers of Mathematics 1980 Agenda for Action that "mathematics programs take full advantage of the power of computers and calculators at all grade levels" is a long way from universal implementation. Availability of computers is spotty with access varying from relatively good in some schools to nonexistent in others. In addition to a shortage of hardware, in many schools there is a shortage of teachers with knowledge of how to effectively use computers in instruction. In the past few years we have seen a proliferation of computer software, yet much of it is of low quality and a high proportion is designed for drill and practice of low level skills. We are still looking to the future for widespread use of computers in a variety of creative ways in mathematics instruction. The potential of the computer for demonstration, problem solving, inquiry, simulation, instructional management, generation of materials and other applications is yet to be realized.

The initiation of the advanced placement exam in computer science in 1984 is another step in the infusion of computer technology into the curriculum. The number of students taking the computer science advanced placement exam is relatively small. We can expect the number to grow in future years, but there are factors that will restrict the growth: the difficulty of the course, the already crowded curriculum, and the lack of qualified teachers. There is a fear that teachers who become well qualified to teach the course will be hired away by industry.

There is a danger that computers, which have so much potential for education, can magnify the inequities that already exist in our society today. Just as inequities in mathematics participation by sex are being reduced, we are witnessing great inequities by sex in computer classes and supplementary uses of computers. Schools in affluent areas tend to have more computers and more teachers with knowledge of computers. Students from more affluent homes are the ones who have computers at home. We can expect to see great diversity in the ability of our student population to use computers; today we have some students who are computer whizzes by the time they leave elementary school and others who still have no computer experience when they leave high school.

The Influence of Tests

Tests can be powerful tools for improving instruction or for perpetuating the status quo. Traditionally, standardized tests tend to follow the trends rather than lead them. While there are new texts on the market every year, the situation is different with standardized norm-referenced tests. There are only a handful of test publishers and the tests are revised infrequently, with a revision cycle that can range from six to ten years. Occasionally we see innovative textbooks that address the needs of a small percent of the national market. However, such is not the case with standardized tests.

In contrast to the standardized norm-referenced tests, the advanced placement tests and scholastic aptitude tests of the Educational Testing Service (ETS) can lead trends. The advanced placement computer science test is an example of such leadership. New tests are prepared by ETS each year and the College Board informs the schools well in advance of any impending significant changes. As the specifics of the project EQuality recommendations are more precisely defined, the College Board tests published by ETS will be instruments for change.

In the near future, the tests that will have the greatest impact will be the ones that are developed at the state and local levels. An example is the screening test developed by Ohio State University which was administered to 61,050 high school juniors throughout Ohio in 1984. After taking the test, the students receive information about which mathematics courses they would be eligible to take at Ohio State or other colleges and universities that participate in the program. Students often conclude as a result of this test that they need to take mathematics during their senior year in high school. Typically when a high school starts using this testing service, enrollment in senior level mathematics courses increases by 40%.

Implications for Two-Year Colleges

As two-year colleges plan mathematics programs they should prepare for a student population that is even more diverse than the population they are serving now. The changes in elementary and secondary mathematics education that are being recommended today will occur unevenly from school to school and from teacher to teacher. The use of computers and calculators in the school curriculum will likely continue to be uneven; therefore, the best plan will be to prepare for everything. As schools place more emphasis on effective mathematics instruction, mathematics achievement of high school graduates may improve. That improvement will probably be accompanied by higher expectations by students for opportunities in careers and for education in two-year colleges and other post secondary institutions.

In this round of educational reform, the federal government is taking a more passive role than it did in the post-Sputnik era. As a result, the focus of change will be more at the state and local level. This local focus has implications for planning by two-year colleges, most of which tend to serve local populations, rather than regional or national populations. Direct contact with the high schools in the local area served will be even more important than an awareness of national trends.

In planning for the future, the two-year colleges can take a reactive role and just wait and see what kinds of students they will receive from the high schools. In many two-year colleges, this approach has been taken in the past; the result has been that approximately 40% of the courses that students take in two-year colleges are courses that they should have successfully completed in high school.

An alternative approach would be to take a proactive role and work directly with the high schools to help students realize what mathematics courses they will need for the various programs offered in the two-year colleges. A National Assessment of Educational Progress survey found that the most influential factor related to students continuing to take mathematics in high school is how useful they think it will be.

High school mathematics teachers tend to appreciate communication with teachers in two-year colleges and other post secondary institutions that their graduates attend. The communication helps them improve their instruction and their ability to help counsel their students. The more closely the mathematics faculties of two-year colleges and high schools work together the more effective the mathematics programs at both levels will be.

REFERENCES

Armstrong, J. and Kahl, S. An Overview of Factors Affecting Women's Participation in Mathematics. Denver: National Assessment of Education Progress, 1979.

Carpenter, T. P., Lindquist, M. M., Matthews, W. and Silver, E. A. "Results of the Third NAEP Mathematics Assessment: Secondary School," The Mathematics Teacher. Vol. 76, No. 9, pp. 652-659. Reston, Va. National Council of Teachers of Mathematics, December, 1983.

Conference Board of The Mathematical Sciences. The Mathematical Sciences Curriculum K-12: What Is Still Fundamental and What Is Not. Washington, D.C.: National Science Foundation, 1983.

Educational Equality Project. Academic Preparation for College: What Students Need To Know and Be Able To Do. New York: The College Board, 1983.

Fey, J. T., Albers, D. J., and Fleming, Wendell H. Undergraduate Mathematical Sciences in Universities, Four-Year Colleges, and Two-Year Colleges, 1980-81. Washington, D.C.: Conference Board of the Mathematical Sciences, 1981.

Gallup, G. H., "The 15th Annual Gallup Poll of the Public's Attitude Toward the Public Schools," Phi Delta Kappa. Vol. 65, No. 1, pp. 33-47. Bloomington, IN: Phi Delta Kappa, September, 1983.

Matthews, W., Carpenter, T. P., Lindquist, M. M. and Silver, E. A. "The Third National Assessment: Minorities and Mathematics." Journal For Research in Mathematics Education. Vol. 15, No. 2. Reston, VA: National Council of Teachers of Mathematics, March 1984.

National Assessment of Educational Progress. Changes in Mathematical Achievement, 1973-78. Denver: Educational Commission of the States, August, 1979.

National Commission on Excellence in Education. "A Nation at Risk: The Imperative for Educational Reform," Education Week (1983): 12-16.

National Council of Supervisors of Mathematics. "Position Paper on Basic Mathematical Skills." The Mathematics Teacher, Vol. 71, No. 2, pp. 147-152. Reston, VA: National Council of Teachers of Mathematics, February, 1978.

National Council of Teachers of Mathematics. An Agenda for Action: Recommendations for School Mathematics of the 1980s. Reston, VA: NCTM, 1980.

U. S. Department of Education. The Nation Responds. Washington, D.C.: U. S. Department of Education, 1984.

(DISCUSSION BEGINS ON P. 471.)

DISCUSSION

COORDINATING CURRICULUM CHANGES

1. Role of State Legislatures

2. Stratification

3. Benign Neglect

4. Expectations

ROLE OF STATE LEGISLATURES

Rodi: One reason why people ought to get interested in getting to know one another (articulation) is that if they don't, then the state legislatures are going to get interested for them, and it would be better to be coming at it from a unified position rather than have legislators coming out and telling you what to do. Texas does have what are called "core curricula"--there has been an agreement in eighteen different areas that wherever a core course is offered at a community college, it must transfer to the four-year school. That doesn't mean that the core courses have to be required for the four-year school degree. But if the four-year school does have college algebra in its degree, and if it's offered at the community college, then that course must be allowed to transfer. The way Texas is doing it is sort of a middle step; they are trying to insure as much transferability as is reasonably possible.

Case: We absolutely cannot at any State University in Florida require, for example, discrete mathematics as an entry to upper division or computer science, until a certain percent of the community college students in the state have a chance to learn it. It doesn't have to be 100%, but of the 28 community colleges, we wouldn't think of doing it until almost all of them that fed us did have it.

Rodi: My point is that if you wanted to change your curriculum, and if there was a tradition of your getting together with at least those community colleges that generally fed you, you wouldn't have the State coming in and saying, "Hey, look, we're concerned about spending money on the duplication of effort and concerned about students not having the most efficient and fiscally sound approach to education. So, if articulation efforts take place, you might diminish the number of places where legislators are going to say, "Hey, you know we don't want to be spending double bucks on education."

Stratification

Page: The two-year colleges are willing to have the four-year people teach in them, and the high schools are willing to have the two-year people teach in them also, but this process does not seem to be reversible. You've got to reverse this and have stratification up and down. One of the ways, of course, to do that is if four-year colleges are serious enough to reserve slots for two-year college people when

they offer visiting professorships.
Now, when you (Rodi) said that there should be interaction, maybe you didn't feel that way, but at least you communicated to me that the two-year people should know what's going on in the four-year schools, but I didn't get the impression you felt the same way about four-year people knowing what's going on in two-year schools. Maybe you don't mean it, but in your paper you're making a case that the four-year people are mostly the mathematicians. And since the two-year people are not the mathematicians, four-year people tend to look down on them. On the other hand, the two-year people are the good instructors. We wear different hats, and the four-year people don't seem to know that. People are different, but they can be treated equally and respected as having differences. I believe that there are two major functions of instruction--one is to transmit knowledge. The first part would be the instruction in the future, and the second part would be the research. The point that I'm trying to make is that neither one is better than the other--we need both.

Rodi: Where there is appropriate overlap, one has to encourage interaction. There is an awful lot of undergraduate instruction that goes on in four-year institutions at the freshman-sophomore level. Four-year faculty, who by and large may have greater interest in other things, nonetheless must have some appropriate degree of interest in freshman-sophomore instruction. On the other hand, community college instructors ought to have some appropriate level of knowledge of what changes have occurred in mathematics over the last five years, what's going on in the profession as a whole, and where their students are going. I suspect that many community college mathematics instructors have lost touch with what their transfer students are going to face when they get to the four-year schools.

Curnutt: It seems to me that it's easier for four-year college people to make an identification with what two-year people do on a daily basis, because so many of the things that four-year college people do are the same things that we do. I think it takes a different kind of effort for two-year college people to really internalize and appreciate what four-year college people do. Let's face it, after teaching college algebra and intermediate algebra for ten years, it gets harder and harder to read those MONTHLY articles. It really takes an effort on the part of two-year college people to stay in their (four-year college) ballgames. And I'd like to put the onus right there--we're the ones who have to take that initiative if we want the bridge to be two-way.

Leitzel: It does seem to me that we need local cooperation but that we need national structure to provide for the interchange of ideas. I think there is an extensive lack of awareness among four-year people about what goes on in two-year colleges. I don't sense the attitudes that you sense; I sense just a general lack of awareness and involvement. We are only now concerned about today and a little bit about tomorrow.

Rodi: Is it sort of benign neglect?

Benign Neglect

Albers: I would like to pick up on your term "benign neglect." I think that it is an appropriate use of that term. We are emerging from a period of tremendous growth in education. Over the last twenty years, two-year colleges grew rapidly. Four-year institutions were experiencing similar growth. Everybody was busy in their own house, keeping the roof on. That took a lot of their energy. We saw double sessions in high schools, and one campus after another being added to community college systems. There is little wonder that there may have been some "benign neglect" of two-year colleges by universities and four-year colleges, and some "benign neglect" of high schools by two-year colleges.

Curnutt: During the first five years that I was in the community colleges, a lot of people just didn't want to listen to people in four-year colleges. Two-year college faculty didn't think that four-year college faculty had anything to say to them. They didn't think that the MAA, or any other professional organization for that matter, would have anything to say to them. This is the situation that still prevails on the part of community college hiring bodies when considering Ph.D.'s for their faculty.

Rodi: I think you're dead right that there was a period in time when two-year colleges tended to isolate themselves and when they said, "We're going to do this unto ourselves." Maybe we're beyond that period. The theme that keeps coming back here is that the next five years is likely to be a very rich and opportune time for two- and four-year faculty to work together.

Taylor: The bottom line on this whole thing is communication, and basically I'm happy to see that because that was the bottom line in my paper. You could throw the rest of it away and say, "Yes, we've got to communicate," which is almost a platitude, but I think the reason that we need to communicate, and it's a virgin and compelling reason, is that when I see that 40% of your time and talents are committed to doing things (teaching high-school level courses) that we should have done, I think that's totally inexcusable on our part. The only way we're going to address that issue is not to continue to pass the problem on to you and let you deal with it, but to work together to address those issues. And basically it is communication that is going to solve the problem. The real question is: What should be in the course? We should also inform the students that they should be getting into those courses and be sure that the content of those courses is important. What is important in mathematics education is whether or not it is thoughtful problem-solving or simple mindless manipulation.

Expectations

Taylor: I really believe in the idea of expectations. We have found with our students that this makes all the difference. We find that with our programs and with our colleagues. I'm optimistic, and I want to end at least this series on articulation on an optimistic note. I'm encouraged that the Sloan Foundation funded this conference and got this group together to address these issues. I believe we will come up with some thing that can go on from here. We appreciate Steve Maurer's coming and representing the Sloan Foundation. When he comes back from the West, he will be known on the East Coast as the "Sloan Ranger."

Let me say why I'm optimistic. At the national level, there has been a very significant increase between the last two national assessments at the 13-year old level, but there hasn't been much increase at the 17-year old level. Those 13-year olds will be the kids who will be coming along in a few years. You will be getting them in the colleges. You can expect improved performance as a result of the pressures put on us by the Nation at Risk report. You will see, I am sure, an improvement. We can make a difference; we can do things.

INDEX